应用型本科院校计算机类专业校企合作实训系列教材

Java EE项目实训教材
——MVC框架技术应用

主　编　杨种学　王小正
副主编　李国青

南京大学出版社

序　言

在当前的信息时代和知识经济时代，计算机科学与信息技术的应用已经渗透到国民生活的方方面面，成为推动社会进步和经济发展的重要引擎。

随着产业进步、学科发展和社会分工的进一步精细化，计算机学科新知识、新领域层出不穷，多学科交叉与融合的计算机学科新形态正逐渐形成。2012年，国家教育部公布的《普通高等学校本科专业目录(2012年)》中将计算机类专业分为计算机科学与技术、软件工程、网络工程、物联网工程、信息安全、数字媒体技术等专业。

随着国家信息化步伐的加快和我国高等教育逐步走向大众化，计算机类专业人才培养不仅在数量的增加上也在质量的提高上对目前的计算机类专业教育提出更为迫切的要求。社会需要计算机类专业的教学内容的更新周期越来越短，相应的，我国计算机类专业教育也将改革的目标与重点聚焦于如何培养能够适应社会经济发展需要的高素质工程应用型人才。

作为应用型地方本科高校，南京晓庄学院计算机类专业在多年实践中，逐步形成了陶行知"教学做合一"思想与国际工程教育理念相融合的独具晓庄特色的工程教育新理念。学生在社会生产实践的"做"中产生专业学习需求和形成专业认同，在"做"中增强实践能力和创新能力，在"做"中生成和创造新知识，在"做"中涵养基本人格和公民意识；同时学生应遵循工程教育理念，标准地"做"，系统地"做"，科学地"做"，创造地"做"。

实训实践环节是应用型本科院校人才培养的重要手段之一，是应用型人才培养目标得以实现的重要保证。当前市场上一些实训实践教材导向性不明显，可操作性不强，系统性不够，与社会生产实际联系不紧密。总体上来说没有形成系列，同一专业的不同实训实践教材重复较多，且教材之间的衔接不够。

《教育部关于"十二五"普通高等教育本科教材建设的若干意见(教高[2011]05号)》要求重视和发挥行业协会和知名企业在教材建设中的作用，鼓励行业协会和企业利用其具有的行业资源和人才优势，开发贴近经济社会实际的教材和高质量的实践教材。南京晓庄学院计算机类专业积极开展校企联合实训实践教材建设工作，与国内多家知名企业共同规划建设"应用型本科院校计算机类专业校企合作实训系列教材"。

本系列教材是在计算机学科和计算机类专业课程体系建设基本成熟的基础上，参考《中国计算机科学与技术学科教程2002》(China Computing Curricula 2002，简称CCC2002)并借鉴ACM和IEEE CC2005课程体系，经过认真的市场调研，由我校优秀教学科研骨干和行业企业专家通力合作而成的，力求充分体现科学性、先进性、工程性。

本系列教材在规划编写过程中体现了如下一些基本组织原则和特点。

1. 贯彻了"大课程观"、"大教学观"和"大工程观"的教学理念。教材内容的组织和案例的甄选充分考虑了复杂工程背景和宏大工程视野下的工程项目组织、实施和管理，注重强化具有团队协作意识、创新精神等优秀人格素养的卓越工程师的培养。

2. 体现了计算机学科发展趋势和技术进步。教材内容适应社会对现代计算机工程人才培养的需求，反映了基本理论和原理的综合应用，反映了教学体系的调整和教学内容的及时更新，注重将有关技术进步的新成果、新应用纳入教材内容，妥善处理了传统知识的继承与现代工程方法的引进。

3. 反映了计算机类专业改革和人才培养需要。教材规划以 2012 年教育部公布的新专业目录为依据，正确把握了计算机类专业教学内容和课程体系的改革方向。在教材内容和编写体系方面注重了学思结合、知行合一和因材施教，强化了以适应社会需要为目标的教学内容改革，由知识本位转向能力本位，体现了知识、能力、素质协调发展的要求。

4. 整合了行业企业的优质技术资源和项目资源。教材采用校企联合开发和建设的模式，充分利用行业专家、企业工程师和项目经理的项目组织、管理、实施经验的优势，将企业实际实施的工程项目分解为若干可独立执行的案例，注重了问题探究、案例讨论、项目参与式教育教学方式方法的运用。

5. 突出了应用型本科院校基本特点。教材内容以适应社会需要为目标，突出"应用型"的基本特色，围绕培养目标，以工程应用为背景，通过理论与实践相结合，重视学生的工程应用能力的培养，增强学生的技能的应用。

相信通过这套"应用型本科院校计算机类专业校企合作实训系列教材"的规划出版，能够在形式上和内容上显著提高我国应用型本科院校计算机类专业实践教材的整体水平，继而提高计算机类专业人才的培养质量，培养出符合经济社会发展需要和产业需求的高素质工程应用型人才。

李洪天

南京晓庄学院　党委书记　教授

前　言

Java EE技术经过多年的发展已日趋成熟。目前,使用基于MVC框架的Java EE技术进行项目开发的企业和工程师越来越多。因此,计算机类专业的学生可以通过掌握Java EE技术来提升就业竞争力。

Java EE技术建立在开源软件基础之上,它是许多软件开发工程师在开发实践中不断摸索提炼出的技术。Java EE技术所包含的知识点非常庞杂,如何指导缺乏软件项目开发经验的学生或初学者较快地理解并掌握Java EE不是一件容易的事。当前,尽管很多高校开设了Java EE课程,但教学效果很不理想,合适教材的缺乏就是原因之一。主要体现在:第一,目前介绍Java EE技术的教材更注重知识的传授,在能力培养方面有所欠缺;第二,教材所举案例比较零散,缺少来源于企业的真实案例,不利于工程实践知识的传授和能力的培养;第三,大多数已有教材包含的知识点对于没有任何开发经验的大学生以及初学者而言难度偏大。因此,编写出符合高校计算机技术和软件工程课程教学特点和需求的教材已刻不容缓。

本书的特点通俗易懂、实用性强以及综合案例来源于企业真实开发项目。编写人员由国内知名企业资深项目架构师和有丰富开发经验的高校教师组成。本书在内容的选取及章节的安排上相对于目前已有的教材做了一定的调整,删减了一些对于初学者而言难以理解的内容。本书前5章内容主要包括:Java EE开发环境配置、JSP应用、Struts 2应用、Hibernate应用、MyBatis应用和Spring应用。每个知识点都从最简单的例子着手,一步一步引导读者学习和实现这些案例。读者在案例实现过程中对涉及的知识点有了初步认识,即先达到"知其然";教材最后两章通过"教学管理系统"和"教育资源网络平台"这两个来源于企业的真实案例使读者进一步掌握和运用所学知识,最后达到"知其所以然"。经过整个课程的系统学习,读者不仅掌握了相关专业知识,并且对企业项目实际的开发流程有了一定的了解。

本书的完成得益于许多老师、学生以及合作企业的积极参与。其中,中兴软件技术有限公司的李国青等工程师提供了部分素材资料,徐家喜、侯青老师以及宇汝军和伊昭荣同学参与了部分章节的案例代码调试和文字内容的编写工作,包依勤和朱杰老师参与了教材的审阅工作。在此一并向他们表示感谢。

本书第1～5章部分内容是对互联网资料进行收集、整理和改编的结果,网站来源包括软件官方网站或软件教程网等。由于本书的大多数知识点和实验内容是建立在开源软件的基础上,而开源软件最大的特点就是广大开源软件爱好者的无私奉献。因此,本人在这里表示对他们的敬意和感谢。

由于作者水平有限,疏漏和错误在所难免,敬请广大师生、读者批评指正。意见和建议可反馈至邮箱:xz_wang@163. com。

本教材提供用于教师教学的配套光盘,光盘包含了教材中所有案例的源代码和开发工具。

编者

2012年12月

目　　录

第1章 Java EE 简介

学习目标

1. 了解 Java EE 体系结构。
2. 了解 Java EE 体系结构的优点。
3. 了解 Java EE 主流的轻型框架。
4. 掌握 JDK、Tomcat 和 MyEclipse 安装配置。

Java 是一种通用、并行、基于类的，且面向对象的程序设计语言，是由 Sun Microsystems 公司于 1995 年 5 月推出的 Java 程序设计语言和 Java 平台(即 Java SE，Java EE，Java ME)的总称。Java 平台由 Java 虚拟机(JVM)和 Java 应用程序接口(API)构成。Java 语言与 C、C++语言有相似之处，也有很大差别。Java 语言略去了 C、C++的一些特性而引入了其他语言的一些思想，Java 语言风格较为接近 C#(C Sharp)。

Java 语言是强类型定义语言，这有助于编程组很快发现问题，因为程序编译时就可以检测出类型错误。Java 语言也是静态语言，通过编译把程序源代码按 JVM 定义规范转换成独立于机器的字节代码，所以 Java 程序无需重新编译便可在不同类型的计算机上执行，即“编写一次，到处运行”。

Java 语言在内存管理、多线程等方面提供了相对简单的管理模式，使得程序员更容易学习和掌握。

综上所述，使用 Java 语言开发更快捷、方便，开发出的软件易于维护与扩展。因此，Java 语言在计算机的各种平台、操作系统，以及手机、移动设备等方面均得到广泛的应用。

§1.1 Java EE 应用概述

目前，Java 平台有三个版本，它们分别是适用于小型设备和智能卡的 Java 平台 Micro 版(Java Micro Edition，Java ME)、适用于桌面系统的 Java 标准版(Java Standard Edition，Java SE)、适用于创建服务器应用程序和服务的 Java 企业版(Java Enterprise Edition，Java EE)。其中 Java EE 是一套开发、部署和管理相关的复杂问题的体系结构，使用 Java 为主要编程语言，减轻企业开发力度，提供一系列模式的解决方案。Java EE 规范是建立在 Java 编程语言和 Java SE 基础之上的规范，是为适应企业的网络需求开发的，保留了标准版的很多优点。

在新的技术需求下，Java EE 作为不同于以往的技术架构，规范了整个系统的开发流程，为开发者们实现了一套全新的技术框架。Java EE 的全称叫做 Java 企业版(Java，Enterprise

Edition)技术规范与指南,里面包含了大量的具体服务架构、组件及技术,这些都是有着共同的标准及规格。Java EE 通过减少 XML 配置和简化 JAR 包以及使用更多的 POJO 和注解的设计方式来提高项目的开发效率。Java EE 依赖于不同商业平台的系统,有了较强的兼容性,为企业系统的过渡、开发、运行提供了强有力的保证。

1.1.1 Java EE 应用的四层结构

(1) 运行在客户端机器上的客户层:负责与用户直接交互。Java EE 支持多种客户端,可以是 Web 浏览器,也可以是专用的 Java 客户端程序。

(2) 运行在 Java EE 服务器上的表示层:该层是基于 Web 的应用服务,利用 Java EE 中的 JSP 与 Servlet 技术,响应客户端的请求,并可向后访问业务逻辑组件。

(3) 运行在 Java EE 服务器上的业务逻辑层:主要封装了业务逻辑,完成复杂计算,提供事务处理、负载均衡、安全、资源连接等各种基本服务。

(4) 运行在 EIS(Enterprise Information System)层服务器上的企业信息系统:该层包括了企业现有系统(数据库系统,文件系统等)。

Java EE 应用的四层结构如图 1-1 所示。

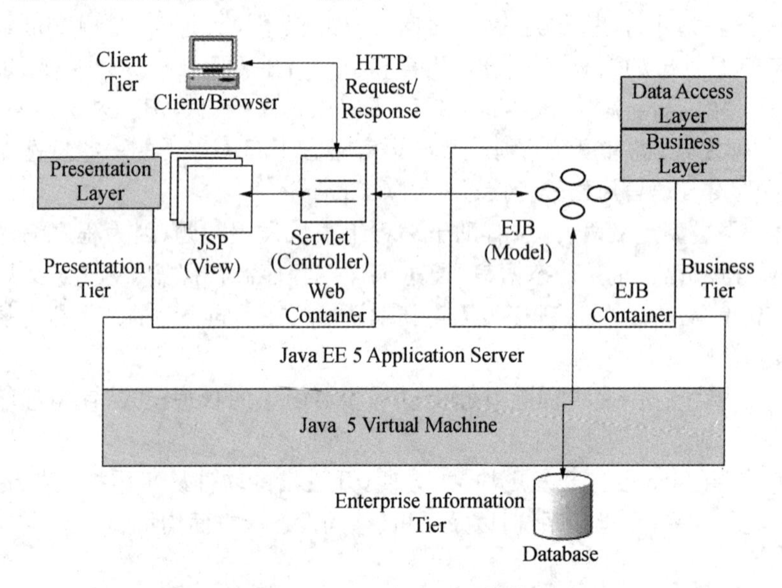

图 1-1 Java EE 四层结构

1.1.2 Java EE 应用的体系结构优点

1. 部署代价廉价

JIEE 体系结构提供中间层集成框架,以满足无需太多费用而又需要高可用性、高可靠性和可扩展性应用的需求。降低了开发多层应用的费用和复杂性,同时提供对现有应用程序集成强有力支持。

2. 开发高效

允许公司把一些通用的、很繁琐的服务端任务交给中间件供应商去完成。这样开发人员可以集中精力在如何创建商业逻辑上,从而大大缩短开发时间。中间件供应商一般提供以下中间件服务:

① 状态管理服务；
② 持续性服务；
③ 分布式共享数据对象 CACHE 服务。

3. 支持异构环境

基于 Java EE 的应用程序不依赖任何特定操作系统、中间件和硬件。Java EE 的程序只需开发一次就可部署到各种平台上，Java EE 标准允许客户订购与 Java EE 兼容的第三方的现成的组件，把它们部署到异构环境中。

4. 可伸缩性

Java EE 平台提供了广泛的负载均衡策略，它能消除系统中的瓶颈，允许多台服务器集成部署。这种部署可达数千个处理器，从而实现高度可伸缩。

§1.2　Java EE 的轻型框架简介

软件开发框架将软件应用中的共性功能抽象出来，预先形成封装好的、与底层无关的及简单易用的接口。框架中通常还集成了很多类库，软件开发人员可以根据需要有选择性地调用或重写，从而完成对数据源、网络、系统等底层构建的访问。另外，软件开发框架并不完全等同于类库。Java EE 框架除了提供类库外，还提供了 IOC（控制反转）的功能。在使用框架开发软件的过程中，对象实例化及方法调用是由框架实现的，其根本目的是缩短开发周期，提高开发效率，提高软件的健壮性和可重用性。

Struts、ORM、Spring、Spring MVC 都是当前 J2EE 开发 Web 应用的主流框架，下面先来简要介绍这四个框架。

1.2.1　Struts 框架

Struts 是一种基于 MVC 经典设计模式的开放源代码的应用框架，也是目前 Web 开发中比较成熟的一种框架。"Struts"的含义即为专业应用开发提供一种"无形的支撑"，它通过把 Servlet、JSP、JavaBean、自定义标签和信息资源等 Java 平台的各种元素整合到一个统一的框架中，为 Web 开发提供具有高可配置性的 MVC 开发模式。

Struts 体系结构实现了 MVC 设计模式的概念，它将 Model、View 和 Controller 分别映射到 Web 应用的组件中。Mode 提供了应用程序的核心功能，它包含应用程序数据和商业逻辑，并且封装了应用程序的状态。View 由 JSP 和 Struts 提供的自定义标签（JSTL、JSF）来实现。Controller 是负责流程控制，充当 Model 和 View 之间的桥梁。由 ActionServlet 负责读取 struts-config. xml 文件，并使用 ActionMapping 来查找对应的 Action。Struts 的体系结构与工作原理如图 1－2 所示。

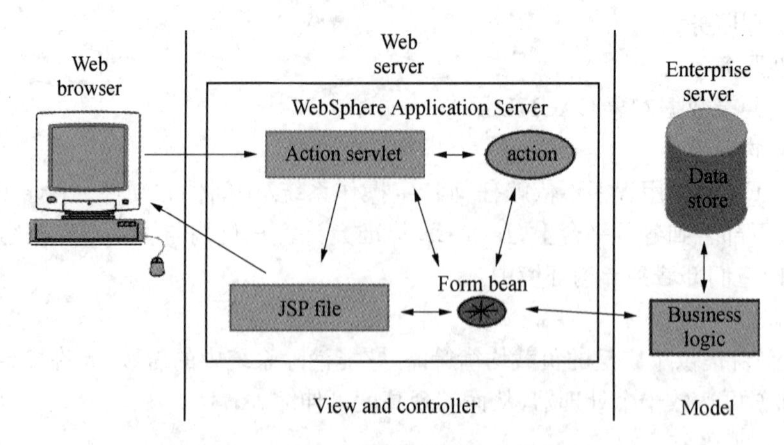

图 1-2　Struts 的体系结构与工作原理

1.2.2　ORM 框架

对目前的 Java EE 信息化系统而言，通常采用面向对象分析和面向对象设计的软件开发流程。系统从需求分析到系统设计都是按面向对象方式进行。通常，在一个程序应用中，都需要传递并持久化对象。传统方法是打开 JDBC 连接，接着创建 SQL 语句并把所需的参数值传递给它。若对象的参数较少时，这样做还比较容易；若对象的参数很多时，实现和维护就很麻烦了。同时，这种实现方法也不符合面向对象的软件开发思想。因此，对象-关系映射(Object-relational mapping)应运而生。

ORM(Object-relational mapping)是一种基于 SQL 模式把对象模型映射到关系型数据模型的数据映射技术。Hibernate 和 iBATIS 是面向 Java 环境的对象/关系映射工具，它可将对象模型表示的对象映射到基于 SQL 的关系数据模型中。ORM 把相应 Java 对象操作自动转换成 SQL 操作，程序开发者可以很容易地持久化 Java 对象。

1.2.3　Spring 框架

Spring 框架是由 Rod Johnson 开发的，2003 年发布了 Spring 框架的第一个版本。Spring 是一个从实际开发中抽取出来的框架，因此，它完成了大量开发中的通用步骤，从而大大提高了企业应用的开发效率。

Spring 为企业应用的开发提供了一个轻量级的解决方案。其中依赖注入、基于 AOP 的声明式事务管理、多种持久层的整合等最受人们关注。Spring 可以贯穿程序的各层之间，能够高效地组织应用程序中的各种中间层组件，但它并不是要取代那些已有的框架(如 Struts、Hibernate 等)，而是以高度的开发性与它们紧密地整合，这也是 Spring 被广泛应用的原因之一。

1.2.4　Spring MVC 框架

Spring MVC 是 Spring Framework 的一个 Web 组件，已经融合在 Spring Web Flow 里面。Spring 框架提供了构建 Web 应用程序的全功能 MVC 模块。使用 Spring 可整合的 MVC 架构，可以选择是使用内置的 Spring Web 框架还是 Struts、JSF 等 Web 框架。通过策略接口，Spring MVC 框架是高度可配置的，Spring MVC 框架并不需要知道使用的视图，它包含多种视图技术，例如 Java Server Pages(JSP)技术、Servlet 和 Titles 等。

Spring MVC 分离了控制器、模型对象、分派器以及处理程序对象的角色，这种分离让它们更容易进行订制。异于同其他 View 框架(Titles 等)无缝集成，采用 IOC 便于测试。

§1.3　JSP 开发环境的搭建

1.3.1　JDK 的安装设置

1. JDK 的下载和安装

JDK(Java Development Kit)即 Java 开发工具包，JSP 的运行环境是基于 JDK 的。

注意：本书 JDK 的版本是 jdk1.7.0_06，可以在"http://java.sun.com"下载文件 jdk-jdk1.7.0_06-windows-i586-p.exe，JDK 的环境变量配置如下。

在"我的电脑"上单击鼠标右键，在弹出的快捷菜单中执行【属性】命令，在弹出的对话框中选择【高级】选项卡，单击【环境变量】按钮，在打开对话框中添加如下的环境变量：

(1) 设置 JAVA_HOME 变量为 Java 的主目录 C:\Program Files\Java\jdk1.7.0_06。

```
$JAVA_HOME= C:\Program Files\Java\jdk1.7.0_06;
```

(2) 把 Java 的 bin 目录路径 D:\ jdk1.5.0\bin 添加到 PATH 环境变量中：

```
$PATH=%JAVA_HOME%\bin;
```

(3) 设置 CLASSPATH 变量：

```
$CLASSPATH= .;%JAVA_HOME%\lib\tools.jar;%JAVA_HOME%\lib;
```

注意：在 CLASSPATH 环境变量的设置中，包含了一个"."的路径，这个"."代表系统的当前路径。如果没有增加该路径，可能导致运行 Java 程序时，class 文件已在当前路径，但在系统提供的文件中找不到该文件。

2. JDK 测试

(1) 用文本编辑器写一个简单的 java 程序：

```
public class HelloWorld {
  public static void main(String args[]) {
      System.out.println("Hello World!");
  }
}
```

这个例子就是著名的"Hello World"，它的功能就是显示"Hello World"。

注意：该文件名称必须为"HelloWorld.java"，大小写也要区分。文件名和 public class 后的名字必须一致。

(2) 编译：在 dos 命令提示符下执行：(注意大小写)

```
javac HelloWorld.java
```

如果运行正常，将生成 HelloWorld. class 文件。

(3) 运行：在 dos 命令提示符下执行：

```
java HelloWorld
```

1.3.2 Tomcat 的安装设置

1. Tomcat 下载

Apache Jakarta 项目组开发的基于 GPL 自由软件协议的 JSP 引擎，配合 JDK 就可以搭建起一个最简单的 JSP 试验平台。在“http://tomcat.apache.org/”下载版本为 7.0.19 的文件 apache-tomcat-7.0.19-windows-x86.zip。

2. 安装 Tomcat

直接解压 apache-tomcat-7.0.19-windows-x86.zip 文件，双击 startup.bat 文件，启动 Tomcat。启动 Tomcat 之后，打开浏览器，在地址栏输入“http://localhost:8080”，然后按回车键，浏览器出现如图 1-3 所示界面，即表示 Tomcat 安装成功。

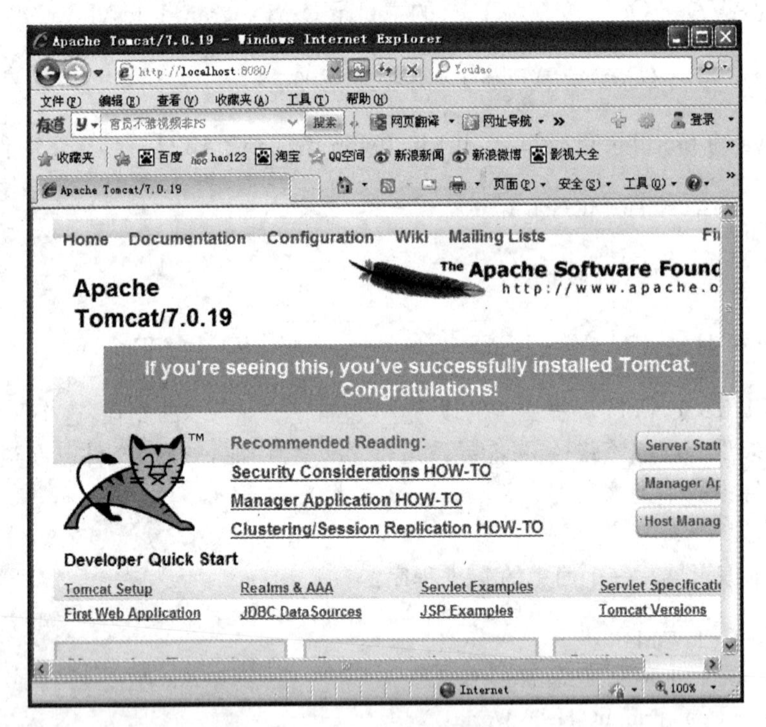

图 1-3 Tomcat 启动

3. Tomcat 基本配置

Tomcat 作为一个 Web 服务器，默认的服务端口是 8080，但该端口完全可以由用户自己控制。虽然 Tomcat 是免费的 Web 服务器，但也提供了两个图形界面的控制台。用户可以使用控制台方便地部署 Web 应用、配置数据源及监控服务器中的 Web 应用等。

下面介绍如何修改 Tomcat 的 Web 服务端口，并进入其控制台来部署 Web 应用。

(1) 修改端口

Tomcat 的配置文件都放在 conf 路径下，控制端口的配置文件也放在该路径下。打开

conf 下的 server. xml 文件，必须使用记事本或 Linux Vim 等无格式的编辑器，不能使用如写字板等有格式的编辑器。在 server. xml 文件中看到如下代码：

```
<Connector port="8080" protocol="HTTP/1.1"
           connectionTimeout="20000"
           redirectPort="8443" />
```

其中 port="8080"，就是 Tomcat 提供 Web 服务的端口。将 8080 修改成任意的端口，建议使用 1000 以上的端口，避免与公用端口冲突。如将服务端口修改为 8117，则 Tomcat 的 Web 服务的提供端口为 8117。修改成功后，重新启动 Tomcat，在浏览器中输入“http://localhost:8117”，按回车键将再次看到如图 1-4 所示的界面，即显示 Tomcat 设置修改成功的界面。

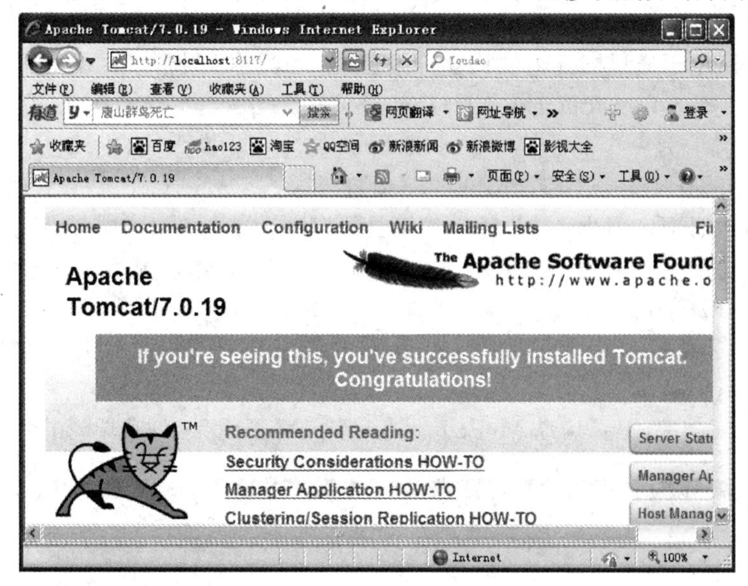

图 1-4　Tomcat 运行

(2) 部署 Web 应用

在 Tomcat 中部署 Web 应用的方式非常多，主要有如下方式：

① 使用控制台部署；

② 利用 Tomcat 的自动部署；

③ 修改 server. xml 文件部署 Web 应用；

④ 增加用户的 Web 部署文件。

通过控制台的部署方式实质上和修改“server. xml”文件的部署方式相同。所有在控制台做的修改，最终都由服务器转变为修改“server. xml”文件。不推荐采用修改“server. xml”文件的配置方式。因为“server. xml”文件是一个系统文件，通常对于系统的文件应尽量避免修改。可通过增加自己的配置文件进行配置。

下面主要介绍用自动部署来增加用户的 Web 部署文件。

自动部署很简单，只需将 Web 应用复制到 Tomcat 的 webapps 路径下，Tomcat 就会自动加载该 Web 应用。增加用户的 Web 部署文件后，为了避免复制 Web 应用，只需简单地增加一个配置文件即可。在 Tomcat 的“cont\Catalina\localhost”路径下，新建一个“. xml”文件，文件可随意命名，但为了更好的可读性，可使该文件的文件名与部署的 Web 应用一致。

其中，每个 Context 元素都对应一个 Web 应用，该元素的 path 属性确定 Web 应用的虚拟路径，而 docBase 则是 Web 应用的文档路径。如“E:/webroot”是一个 Web 应用，若想将该应用部署在/test 虚拟路径下，只需将该文件的内容做如下修改：

```
<! -- 部署一个 Web 应用，其中 path 是 Web 应用虚拟路径，而 docBase 是 Web 应用的文档路径--> 
    <Context path="/test" docBase="e:/webroot" debug="0" privileged ="true">
    </Context>
```

1.3.3 MyEclipse 的安装与设置

Eclipse 是一个开放源代码的、基于 Java 的可扩展开发平台。就其本身而言，它只是一个框架和一组服务，用于通过插件组件构建开发环境。Eclipse 附带了一个标准的插件集，包括 Java 开发工具（Java Development Kit，JDK）。MyEclipse 是一个用于开发 Java 应用的 Eclipse 插件集合，它是一个商业软件。MyEclipse 的功能非常强大，支持也十分广泛，尤其是对各种开源产品的支持十分不错。MyEclipse 目前支持 Java Servlet、AJAX、JSP、JSF、Struts、Spring、Hibernate、EJB3、JDBC 数据库链接工具等多项功能。对于初学者来说，MyEclipse 更易上手，因此本书使用 MyEclipse 作为 Java EE 开发平台，在本书最后一章的综合项目中使用了 Eclipse 作为开发平台。

MyEclipse 的安装过程非常简单，这里不再详细列举，本书示例用的是完全安装的 MyEclipse。

MyEclipse 安装完成后，启动 MyEclipse，出现了 MyEclipse 的菜单，如图 1－5 所示。

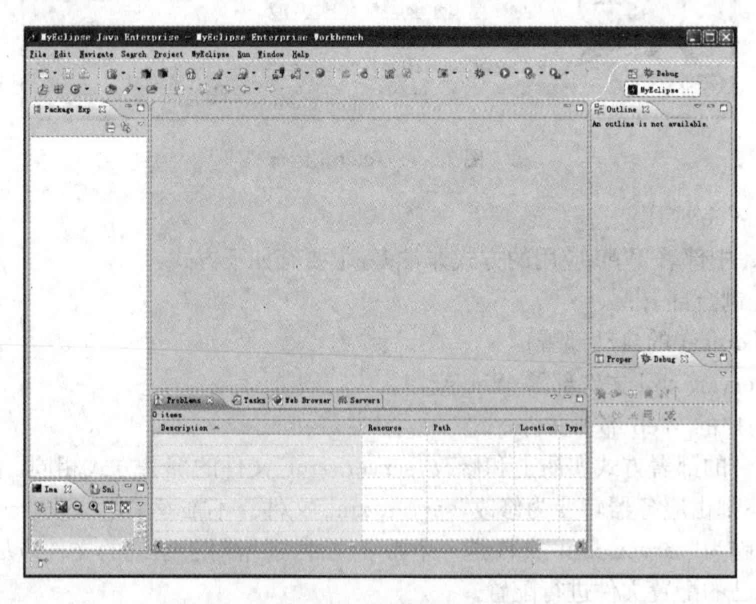

图 1－5 MyEclipse 运行

现在对 MyEclipse 的 Tomcat 进行配置。选择【Window】→【Preferences】→【MyEclipse】→【Servers】→【Tomcat】菜单项，选择【Tomcat 6. x】，在右边的【Tomcat server】栏中选择【Enable】，单击【Tomcat home directory】后面的【Browser】按钮，选择 Tomcat 的安装路径“E:\software\apache-tomcat-7.0.19”，下面的两行会自动生成出来，不做修改，如图 1－6 所示。

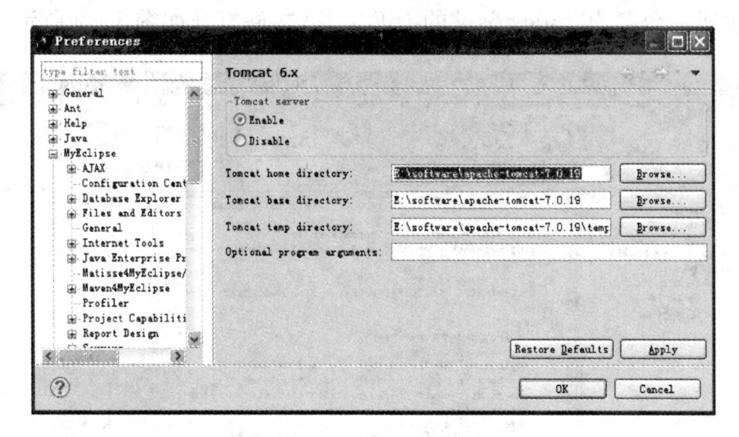

图 1－6　MyEclipse 配置 Tomcat

单击【OK】按钮，再查看菜单栏的服务器配置，若有“Tomcat 6.x”则说明配置成功，如图 1－7所示。由于 MyEclipse 自身带有 JDK，所以不用另外配置。

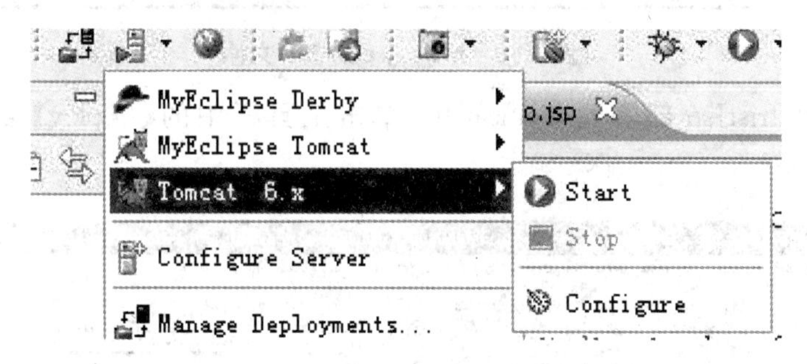

图 1－7　MyEclipse 配置 Tomcat 成功

§1.4　应用实例

打开 MyEclipse 程序，新建一个 Web Project，如图 1－8 所示。

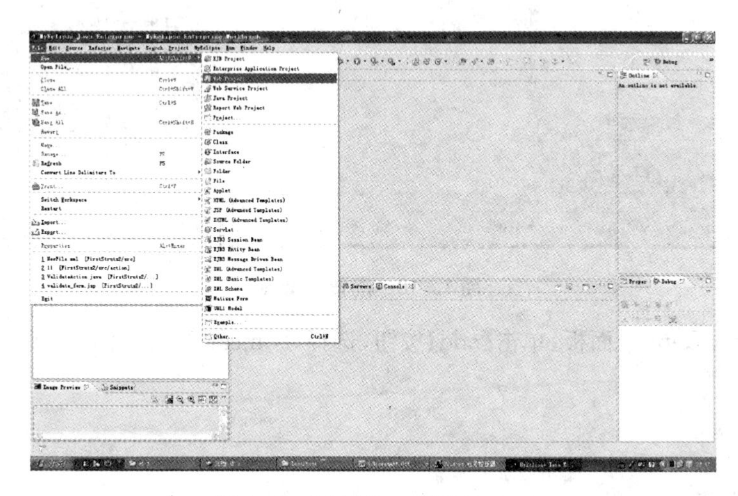

图 1－8　MyEclipse 新建 Project

程序自动生成一个名为“index. jsp”的 JSP 文件，文件打开如图 1 - 9 所示。

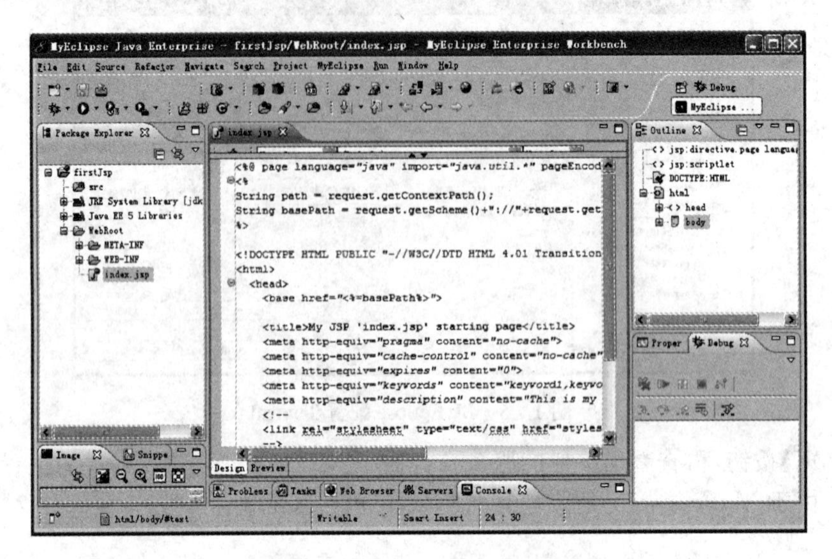

图 1 - 9 index. jsp 的 JSP 文件

接着把项目 fristJsp 部署到 Tomcat 中，单击工具栏中的【Deploy】按钮，如图 1 - 10 所示。

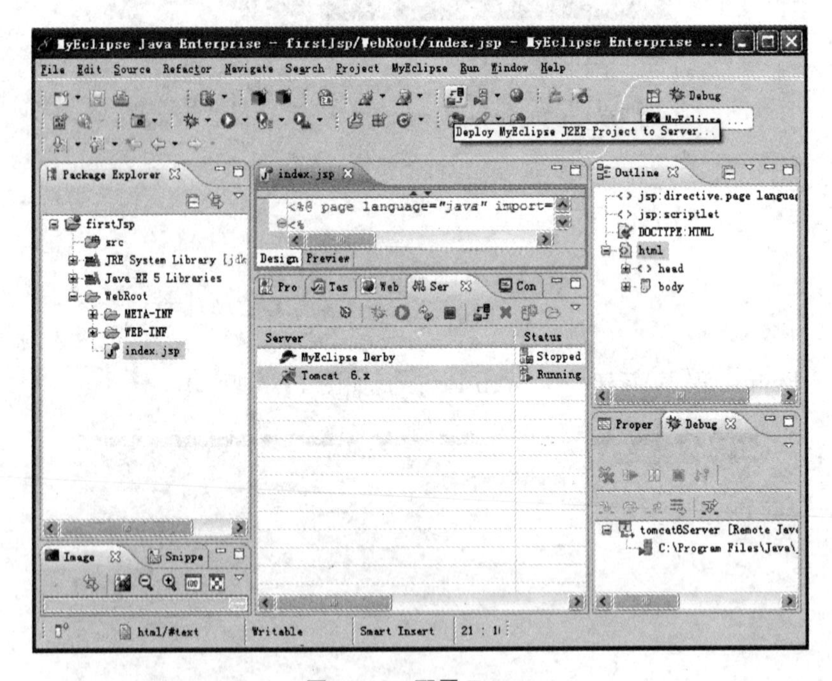

图 1 - 10 配置 Tomcat

打开 Project Deploy 面板，单击【add】按钮，选择 Tomcat 6. x，如图 1 - 11 所示。

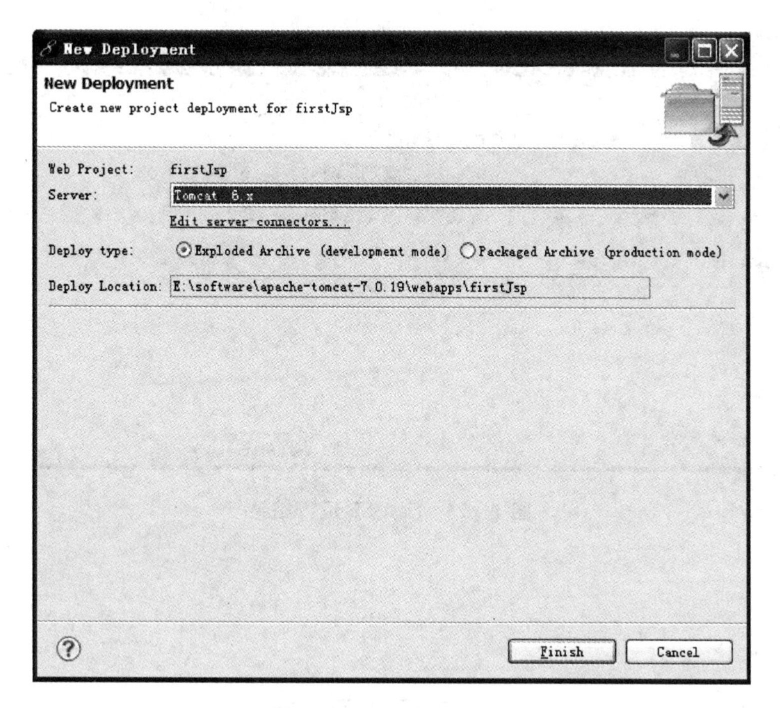

图 1-11 Project Deploy

接着,在 Servers 面板中启动 Tomcat 6. x,如图 1-12 所示。

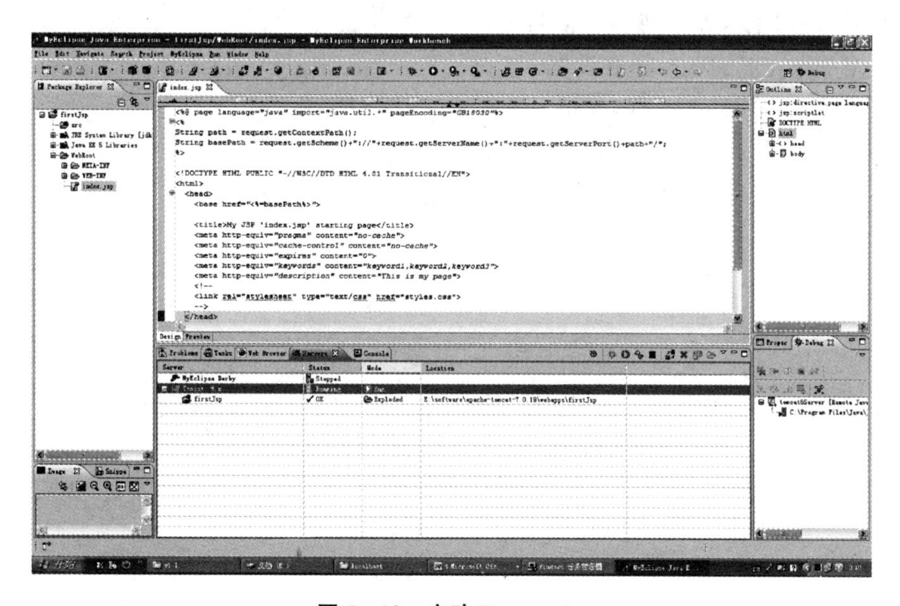

图 1-12 启动 Tomcat 6. x

最终,打开 IE 浏览器,在地址栏输入"http://localhost:8080/firstJsp",看到如图 1-13 所示结果,表示项目部署运行成功。

图 1－13　firstJSP 运行结果

巩固练习

1. 根据不同应用领域，Java 语言可分为哪三大平台？
2. 简述 Java EE 主流开发框架的特点。
3. 下载安装 JDK、Tomcat 并设置。
4. 通过网络学习 Tomcat 的虚拟目录配置的不同方法。
5. 掌握 MyEclipse 中配置 Tomcat 和 JDK。

第 2 章　Java web 编程基础

学习目标

1. 了解并掌握 HTML 的元素使用。
2. 了解 JSP 程序的工作原理。
3. 利用 JSP、JavaBean 开发简单系统。

§2.1　HTML 语言

2.1.1　HTML 基本概念

HTML 的英文全称是 Hyper Text MarkUp Language，中文叫做“超文本标记语言”。它和一般文本的不同的是：一个 HTML 文件不仅包含文本内容，还包含一些 Tag，中文称“标记”。一个 HTML 文件的后缀名是“.htm”或“.html”。用文本编辑器就可以编写 HTML 文件。

HTML 示例：

```
<HTML>
    <HEAD>
        <TITLE>一个简单的 HTML 示例</TITLE>
    </HEAD>
    <BODY>
        <CENTER>
            <H3>欢迎光临</H3>
            <BR>
            <HR>
        <FONT SIZE=2>HTML 文本结构的展示！</FONT>
        </CENTER>
    </BODY>
</HTML>
```

文件的显示效果如图 2－1 所示。

图 2-1 HTML 示例

2.1.2 HTML 的标签

(1) <html>标识文档的开始,</html>标识文档的结束,必须成对使用。

(2) <head>和</head>构成开头,标志之间可使用<title></title>、<script></script>等标志,这些标志对都是描述 HTML 文档相关信息的标志对,之间内容不会显示出来。

(3) <body></body>是 HTML 文档的主体部分,在此标志对之间可包含<p>、</p>、<h1>、</h1>、
、<hr>等众多的标志,它们所定义的内容会显示出来。

(4) <title></title>是浏览器窗口左上角蓝色部分的文本信息,其只能放在<head></head>标志对之间。

2.1.3 表格与框架

1. 表格的基本结构

(1) <table>...</table> 定义表格

(2) <caption>...</caption> 定义标题

(3) <tr> 定义表行

(4) <th> 定义表头

(5) <td> 定义表元(表格的具体数据)

2. 表格举例

```
<html>
    <head>
        <table><title>
    </head>
```

```
        <body>
            <table border="2" bgcolor="aqua">
                <tr>
            <th bgcolor="ffaa00">彩电</th>
            <th bgcolor="Red">冰箱</th>
            <th rowspan="2">家电</th>
                </tr>
                <tr bgcolor="yellow">
                    <td>A</td><td>B</td>
                </tr>
            </table>
        </body>
</html>
```

效果如图 2－2 所示。

图 2－2　HTML 表格举例图

3. 表单的常用控件(如表 2－1 所示)

表 2－1　表单常用控件

表单控件(Form Controls)	说　明
input type="text"	单行文本输入框
input type="submit"	将表单(Form)里的信息提交给表单里 action 所指向的文件
input type="checkbox"	复选框
input type="radio"	单选框
select	下拉框
textarea	多行文本输入框
input type="password"	密码输入框(输入的文字用 * 表示)

4. 表单举例

```
<html>
    <head><title>输入用户姓名</title></head>
    <body>
        <form action="http://www.blabla.cn/asdocs/html_tutorials/yourname.asp" method="
get">
            请输入你的姓名：
            <input type="text" name="yourname"/>
            <input type="submit" value="提交"/>
        </form>
    </body>
    </html>
```

效果如图 2-3 所示。

图 2-3 Html 表单

§2.2 Servlet 与 JSP 简介

2.2.1 Servlet 简介

1. Servlet 简介

Servlet(Server Applet)全称 Java Servlet，是用 Java 编写的服务器端程序。其主要功能是交互式地浏览和修改数据，生成动态 Web 内容。狭义的 Servlet 是指 Java 语言实现的一个接口，广义的 Servlet 是指任何实现了这个 Servlet 接口的类。一般情况下，人们将 Servlet 理解为后者。

Servlet 运行于支持 Java 的应用服务器中。理论上，Servlet 可以响应任何类型的请求，但绝大多数情况下 Servlet 只用来扩展基于 HTTP 协议的 Web 服务器。最早支持 Servlet 标准的是 JavaSoft 的 Java Web Server。此后，一些其他的基于 Java 的 Web 服务器开始支持标准的 Servlet。

2. Servlet 与 Applet 比较

(1) 相似之处：

① 不能独立运行，没有 main()方法。

② 都是由 Servlet 容器调用。

③ 都有一个生存周期，包含 init()和 destroy()方法。

(2) 不同之处：

① Applet 可以有图形界面(AWT)，与浏览器一起，在客户端运行。

② Servlet 则没有图形界面，在服务器端运行。

3. Servlet 的优点

(1) 高效：在 Servlet 中，每个请求由一个轻量级的 Java 线程处理(而不是重量级的操作系统进程)，因此开销很小。如果有 *N* 个对同一 Servlet 程序的并发请求，对于 Servlet 而言处理请求的是 *N* 个线程，因此内存中只保留一个 Servlet 实例。

(2) 方便：Servlet 提供了大量的实用工具例程。例如自动地解析和解码 HTML 表单数据、读取和设置 HTTP 头、处理 Cookie、跟踪会话状态等。

(3) 功能强：在 Servlet 中能够在各个程序之间共享数据，使得数据库连接池之类的功能很容易实现。

(4) 可移植性好：Servlet 用 Java 编写，Servlet API 具有完善的标准。几乎所有的主流服务器都直接或通过插件支持 Servlet。

4. Servlet 与 JSP(JavaServer Pages) 的比较

JavaServer Pages(JSP)是一种实现普通静态 HTML 和动态 HTML 混合编码的技术，JSP 并没有增加任何本质上不能用 Servlet 实现的功能。但是，在 JSP 中编写静态 HTML 更加方便，不必再用 println 语句来输出每一行 HTML 代码。更重要的是，借助内容和外观的分离，页面制作中不同性质的任务可以方便地分开。如 HTML 内容由页面设计者进行设计，而动态程序由 Servlet 程序员进行开发。

5. servlet 简单实例

现在来创建一个简单的 Servlet：FirstServlet 类，功能只是输出“Hello 大家好!”。

首先打开安装了 MyEclipse 插件的 Eclipse，然后建一个 Web 项目。选择【File】→【New】→【Project…】菜单项，弹出新建命令对话框，选择【MyEclipse】→【Java Enterprise Projects】→【Web Project】菜单项，单击【Next】按钮进入 Web 应用详细信息设置，在【Project Name】文本框中输入 Web 应用名称，命名为“FirstServlet”，在【J2EE Specification Level】一栏中选择【Java EE 4.0】菜单项，其他为默认值，如图 2-4 所示，单击【Finish】按钮完成。

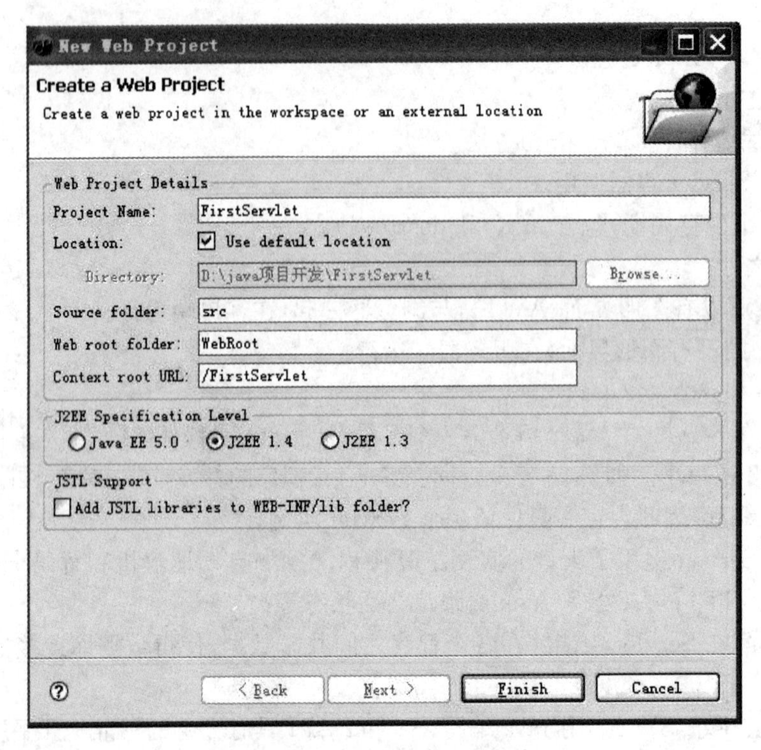

图 2-4 建立 Web 项目

项目建立完成后,在左边的视图中可以看到刚才新建项目的内容,如图 2-5 所示。

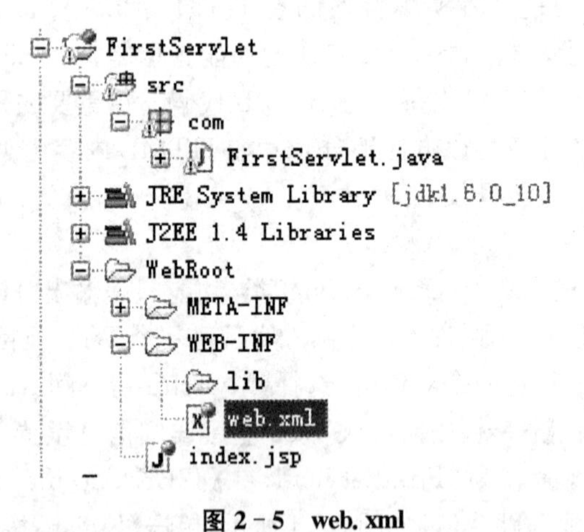

图 2-5 web.xml

右击 src 文件夹,选择【new】→【class】菜单项,弹出新建类对话框,在【name】一栏中输入类名,命名为"FirstServlet",其他为默认值,单击【Finish】按钮完成。

FirstServlet.java 内容如下:

```
package com;
import java.io.IOException;
import java.io.PrintWriter;
import javax.servlet.ServletException;
import javax.servlet.http.HttpServlet;
import javax.servlet.http.HttpServletRequest;
import javax.servlet.http.HttpServletResponse;
public class FirstServlet extends HttpServlet {
    protected void doGet(HttpServletRequest req, HttpServletResponse resp)
            throws ServletException, IOException {
        // 设定内容类型为 HTML 网页 UTF-8 编码
        resp.setContentType("text/html;charset=UTF-8");
        // 输出页面
        PrintWriter out = resp.getWriter();
        out.println("<html><head>");
        out.println("<title>First Servlet Hello</title>");
        out.println("</head><body>");
        out.println("First Servlet!");
        out.println("</body></html>");
        out.close();
    }
}
```

在这段代码中，首先导入“javax.servlet.*”和“javax.servlet.http.*”。其中“javax.servlet.*”存放了与 HTTP 协议无关的一般性 Servlet 类；“javax.servlet.http.*”增加了与 HTTP 协议有关的功能。

所有 Servlet 都必须实现“javax.servlet.Servlet”接口，但通常会从“javax.servlet.GenericServlet”或“javax.servlet.http.HttpServlet”选择其中一个来实现。如果写的 Servlet 代码和 HTTP 协议无关，那么必须继承 GenericServlet 类。若有关，则必须继承 HttpServlet 类。例子中继承的是 HttpServlet 类。

“javax.servlet.*”里面的 ServletRequest 和 ServletResponse 接口提供存取一般的请求和响应；而“javax.servlet.http.*”里面的 HttpServletRequest 和 HttpServletResponse 接口，则提供 HTTP 请求及响应的存取服务。上面代码中实现的是 HttpServletRequest 和 HttpServletResponse 接口。

上面代码中，利用 HttpServletResponse 接口的 setContentType()方法来设定内容类型。本例要显示为 HTML 网页类型，因此，内容类型设为“text/html”，这是 HTML 网页的标准 MIME 类型值。接着，用 getWriter()方法返回 PrintWriter 类型的 out 对象，它与 PrintStream 类似，但是它能够对 Java 的 Unicode 字符进行编码转换。最后，利用 out 对象把“Hello 大家好!”的字符串显示在网页上。

接下来，需要设定 web.xml 文件，web.xml 文件在 Web 项目的 WEB-INF 文件夹内。如上图 2-5 所示。

Web.xml 配置内容如下：

```
<? xml version="1.0" encoding="UTF-8"? >
<web-app version="2.4"
     xmlns="http://java.sun.com/xml/ns/j2ee"
     xmlns:xsi="http://www.w3.org/2001/XMLSchema-instance"
     xsi:schemaLocation="http://java.sun.com/xml/ns/j2ee
     http://java.sun.com/xml/ns/j2ee/web-app_2_4.xsd">
  <servlet>
     <servlet-name> FirstServlet</servlet-name>
     <servlet-class> com.FirstServlet</servlet-class>
  </servlet>
  <servlet-mapping>
     <servlet-name> FirstServlet</servlet-name>
     <url-pattern>/ FirstServlet</url-pattern>
  </servlet-mapping>
</web-app>
```

研究 web.xml 文件中关于 Servlet 的配置了解配置一个 Servlet 需要配置两个标签：一个是<servlet>；另一个是<servlet-mapping>。

在<servlet>标签中，可以配置 Servlet 的名字、调用的 Java 类以及 Servlet 初始化时传入的参数。在本例中，Servlet 名字是"FirstServlet"，调用的 java 类是"com.FirstServlet"，即是 Servlet 的 package 加上类名。在这个简单的 Servlet 中，不需要传递初始化参数给 Servlet，所以没有配置初始化参数，关于配置初始化参数，会在后面的例子里做进一步讲解。

对于<servlet-mapping>，首先指定了 Servlet 的名字，然后设置 Url 连接，Url 设置是"/FirstServlet"。Servlet 名字必须和上面的<servlet>标签中的<servlet-name>的值一致。

当页面中设定的连接和<url-pattern>中设定的值一致时，则会通过<servlet-name>找到对应 Servlet 类来运行。在本例中，当页面的连接（a 标签或 form 设定的 action）是"/FirstServlet"时，则会通过 Servlet 的名字"FirstServlet"来找到对应的 Servlet 类"com.FirstServlet"来运行。

最后，在浏览器中输入网址："http://localhost/Servlet/FirstServlet"，FirstServlet 的执行结果如图 2-6 所示。

图 2-6　运行结果

提示：若 Tomcat 中设置的端口号不是 80，则需要加上端口号。比如 Tomcat 设置的端口号为 8088，则访问网址为"http://localhost:8088/Servlet/FirstServlet"。

6. Servlet 的生命周期

Servlet 的生命周期如图 2-7 所示。

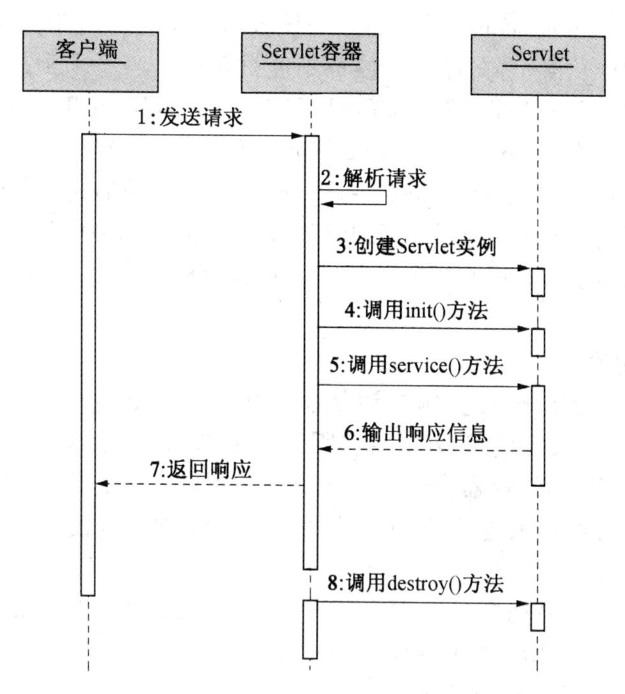

图 2-7 Servlet 的生命周期

Servlet 运行在 Servlet 容器中，其生命周期由容器来管理。Servlet 的生命周期通过"javax. servlet. Servlet"接口中的 init()、service()和 destroy()方法来表示。

Servlet 的生命周期包含了下面四个阶段：

(1) 加载和实例化阶段

Servlet 容器负责加载和实例化 Servlet。当 Servlet 容器启动时，检测到需要 Servlet 来响应请求时，将加载 Servlet。当 Servlet 容器加载 Servlet 类时，必须知道所需的 Servlet 类在什么位置，Servlet 容器可以从本地文件系统、远程文件系统或者其他的网络服务中通过类加载器加载 Servlet 类。Servlet 类成功加载后，容器将创建 Servlet 的实例。因为容器是通过 Java 的反射 API 来创建 Servlet 实例，调用的是 Servlet 的默认构造方法(即不带参数的构造方法)，所以在编写 Servlet 类的时候，不应该提供带参数的构造方法。

(2) 初始化阶段

Servlet 实例化之后，容器将调用 Servlet 的 init()方法初始化这个对象。初始化的目的是为了让 Servlet 对象在处理客户端请求前完成一些初始化的工作，如建立数据库的连接，获取配置信息等。对于每一个 Servlet 实例，init()方法只被调用一次。在初始化期间，Servlet 实例可以使用容器为它准备的 ServletConfig 对象从 Web 应用程序的配置信息(在 web. xml 中配置)中获取初始化的参数信息。这样 servlet 的实例就可以把与容器相关的配置数据保存起来供以后使用。在初始化期间，如果发生错误，Servlet 实例可以抛出 ServletException 异常，一旦抛出该异常，servlet 就不再执行，而随后对它的调用会导致容器对它重新载入并再次运行此方法。

(3) 请求处理阶段

在成功执行 init()方法后，Servlet 容器调用 Servlet 的 service()方法对请求进行处理。在 service()方法中，通过 ServletRequest 对象得到客户端的相关信息和请求信息，在对请求进行处理后，调用 ServletResponse 对象的方法设置响应信息。对于 HttpServlet 类，该方法作为 HTTP 请求的分发器，它在任何时候都不能被重载。当请求到来时，service()方法决定请求的类型（如 GET、POST、HEAD、OPTIONS、DELETE、PUT、TRACE 等），并把请求分发给相应的处理方法（如 doGet()、doPost()、doHead()、doOptions()、doDelete()、doPut()、doTrace()等），每个 do 方法具有和第一个 service()相同的形式。常用方法是 doGet()和 doPost()方法，为了响应特定类型的 HTTP 请求，必须重载相应的 do 方法。如果 Servlet 收到一个 HTTP 请求而没有重载相应的 do 方法，它就返回一个此方法对本资源不可用的标准 HTTP 错误。

(4) 服务终止阶段

当容器检测到一个 Servlet 实例应该从服务中被移除的时候，容器就会调用实例的 destroy()方法，以便让该实例可以释放它所使用的资源，保存数据到持久存储设备中。在 destroy()方法调用之后，容器会释放这个 Servlet 实例，该实例随后会被 Java 的垃圾收集器所回收。如果再次需要这个 Servlet 处理请求，Servlet 容器会创建一个新的 Servlet 实例。

在整个 Servlet 的生命周期过程中，创建 Servlet 实例、调用实例的 init()和 destroy()方法都只进行一次，当初始化完成后，Servlet 容器会将该实例保存在内存中，通过调用它的 service()方法，为接收到的请求服务。

2.2.2　JSP 简介

JSP(Java Server Pages)是由 Sun Microsystems 公司倡导、许多公司参与一起建立的一种动态网页技术标准。JSP 技术有点类似 ASP 技术，它是在传统的网页 HTML 文件(*.htm，*.html)中插入 Java 程序段(Scriptlet)和 JSP 标记(tag)，从而形成 JSP 文件(*.jsp)。用 JSP 开发的 Web 应用是跨平台的，既能在 Linux 下运行，也能在其他操作系统上运行。

JSP 与 Java Servlet 一样，是在服务器端执行的，通常返回给客户端的就是一个 HTML 文本，因此客户端只要有浏览器就能浏览。

JSP 页面由 HTML 代码和嵌入其中的 Java 代码所组成。服务器在页面被客户端请求以后对这些 Java 代码进行处理，然后将生成的 HTML 页面返回给客户端的浏览器。Java Servlet 是 JSP 的技术基础，而且大型的 Web 应用程序的开发需要 Java Servlet 和 JSP 配合才能完成。JSP 具备了 Java 技术的简单易用、完全的面向对象、具有平台无关性且安全可靠以及主要面向因特网的所有特点。

JSP 可用一种简单易懂的等式表示为：HTML＋Java＋JSP 标记＝JSP。

2.2.3　JSP 与 Servlet 的关系

Servlet 是服务器端的程序，动态生成 Html 页面发到客户端，但是这样程序里有许多 out.println()，Java和 HTML 语言混在一起很乱，所以后来推出了 JSP。其实 JSP 就是 Servlet，每一个 JSP 在第一次运行时被转换成 Servlet 文件，再编译成“.class”来运行。有了 JSP，在 MVC 模式中 Servlet 不再负责生成 Html 页面，转而担任控制程序逻辑的作用，控制 JSP 和 JavaBean 之间的流转。图 2－8 描述了 JSP 转化为 Servlet 的过程。

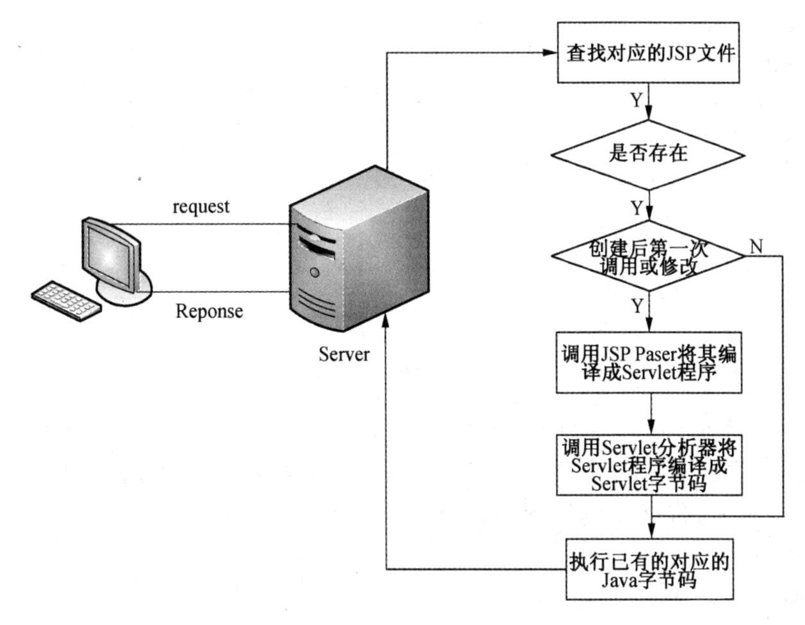

图 2－8　JSP 转化 Servlet 过程

随着 JSP 的广泛应用和各种设计模式的盛行，JSP 也暴露了大量的问题：首先，夹杂服务端代码的 JSP 文件给后期维护和页面风格的再设计带来大量阻碍，美工在修改页面的时候不得不面对大量看不懂的服务端代码，程序员在修改逻辑时常会被复杂的客户端代码搞昏。交叉的工作流使得 JSP 面临大量的困境，直接导致了 Java Web 框架技术的出现，框架技术倡导了 MVC 的概念，这将在后续章节中，详细阐述。

§2.3　JSP 具体内容

2.3.1　JSP 特点

JSP 结合了 Servlet 和 JavaBean 技术，充分继承了 Java 的众多优势，具有以下特点：

① 一次编写，随处运行；

② 组件重用；

③ 页面开发标记化；

④ 角色分离；

⑤ 开发设计。

2.3.2　JSP 基本语法

1. JSP 程序段

(1) 在 JSP 中符合 Java 语言规范的程序被称为程序段。

(2) 程序段包括在“＜％　％＞”之间，基本语法为：＜％code fragment％＞。

例如：

```
<html>
    <head>
      </head>
        <body>
            <%
              int h=10,w=5,s;
              s=h * w;
              out.print(s);
            %>
        </body>
</html>
```

将项目部署在 Tomcat 上，并启动 Tomcat。在浏览器中输入网址“http://localhost:8080/test/first.jsp”，按回车键得到结果显示为“50”。

2. JSP 注释

(1) 普通 Java 注释

① 用“//”注释单行。

② 用“/*　*/”注释多行。

③ 用“/**　*/”注释多行，用于将所注释的内容文档化。

(2) JSP 特有的注释

① 客户端注释，如果使用者检视网页的原始码，也会看到这些注释，语法如下：

```
<! -- comment|<%=expression%>|-->
```

例如：

```
<! --<%=new Java.util.Date()%>-->
```

效果如图 2-9 所示。

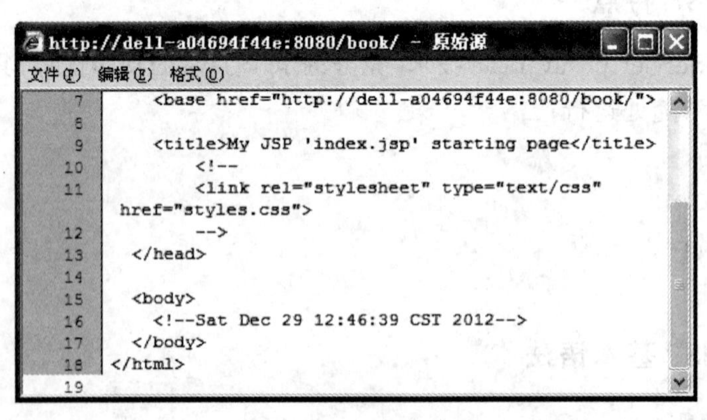

图 2-9　客户端注释

(2) 服务器端注释，使用者在网页原始码中看不到注释，可以将它放在<%-- ? --%>卷标里，语法如下：

```
<%/* comment */%>或<%-- comment --%>
```

例如：

```
<%--=new java.util.Date()--%>
```

效果如图2-10所示。

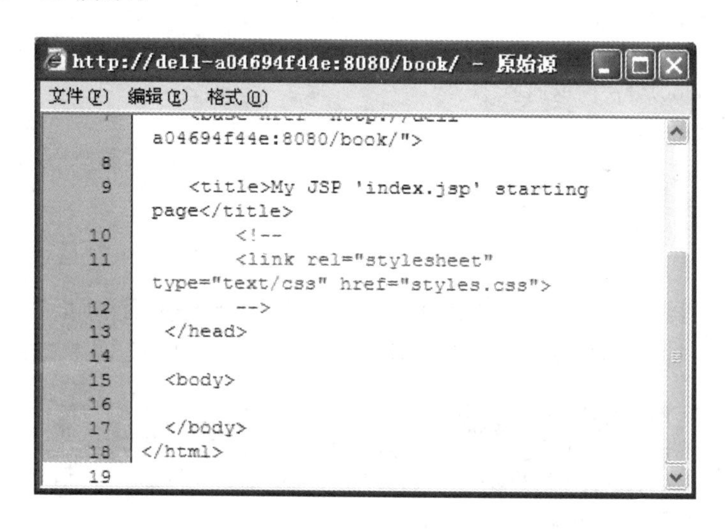

图2-10 服务器端注释

3. JSP声明

JSP声明用于声明变量和方法。在JSP声明中声明方法看起来很特别，没有类，只有方法定义，方法可以脱离类独立存在。实际上，JSP声明将会转换成Servlet的成员变量或成员方法，就是声明一个类的成员变量，因此JSP声明依然符合Java语法。

(1) 使用声明来定义需要使用的变量、方法。

(2) JSP声明方式与Java相同，其语法格式为：

```
<%! declaration;[ declaration;]...... %>
```

例如：

```
<%@ page contentType="text/html; charset=gb2312" language="java" %>
<!DOCTYPE HTML PUBLIC "-//W3C//DTD HTML 4.0 Transitional//EN">
<HTML>
  <HEAD>
  <TITLE>声明测试</TITLE>
  </HEAD>
  <! -- 下面是JSP声明部分 -->
  <%!
    //声明一个整型变量
    public int count;
    //声明一个方法
```

```
        public String info(){return "hello";}
      %>
    <BODY>
      <%
        //将 count 的值输出后再加 1
        out.println(count++);
      %>   <br>
      <%
        //输出 info()方法的返回值
        out.println(info());
      %>
    </BODY>
</HTML>
```

打开多个浏览器，甚至可以在不同的机器上打开浏览器来刷新该页面，将发现所有客户端访问的 count 值是连续的，即所有客户端共享了同一个 count 变量。效果如图2－11所示。

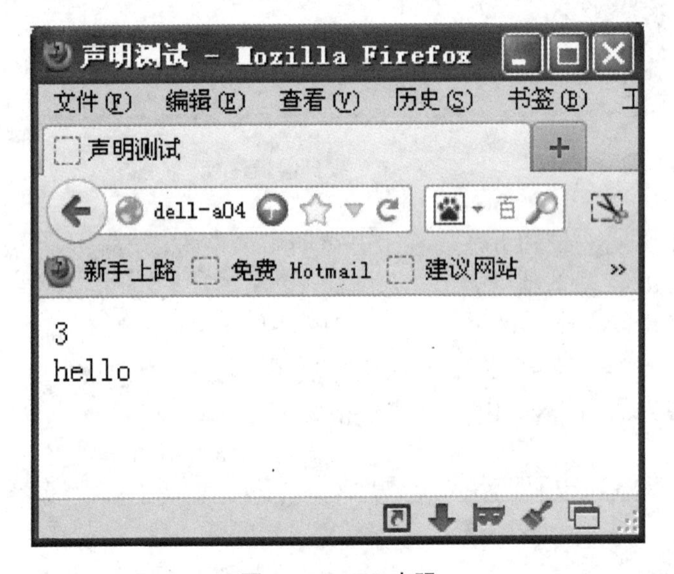

图 2－11 JSP 声明

结果表明：JSP 页面会编译成一个 Servlet 类，每个 Servlet 在容器中只有一个实例；在 JSP 中声明的变量是成员变量，成员变量只在创建实例时初始化，该变量的值将一直保存，直到实例销毁。

4. JSP 表达式

(1) 表达式元素是指在脚本语言中被定义的表达式，其在运行后自动地转化为字符串并显示在浏览器中。

(2) 基本语法为：<%=expression%>

```
<%=new java.util.Date()%>
```

效果如图 2－12 所示。

图 2-12　JSP 表达式

5. JSP 指令标记

用来设置与整个 JSP 页面相关的属性，它并不直接产生任何可见的输出，而是告诉引擎如何处理其余 JSP 页面。

6. page 指令

用来设置整个页面的相关属性和功能，作用于整个页面。

例：

```
<%@ page language="java" import="java.util.*" pageEncoding="GB18030"%>
```

(1) include 指令

用于解决重复性页面问题，其中包含的文件在本页面编译时被引入。

语法：<%@ include file="url" %>

include 指令举例：

```
<%@ include file="top.jsp" %>
```

(2) taglib 指令

用于提供类似于 XML 中的自定义新标记的功能。

语法：<%@ taglib url="relative taglibURL" prefix="taglibPrefix" %>

taglib 指令举例：

首先，声明在 JSP 中使用标签库：

```
<%@taglib prefix="s" uri="/struts-tags"%>
```

其中，uri 属性引用了标签库描述符“TLD”，prefix 属性定义了区别其他标签的前缀名。

接着，可以使用所引用的标签库中的“if”标签，例：

```
<s:if test='#request.teacherInfo.degree==NULL'>
        <option value=" "></option>
</s:if>
```

7. JSP 动作元素

(1) <jsp:include>

① 作用:在当前页面添加动态和静态的资源。

② 基本语法为:<jsp:include page="url"/>。

③ include 指令和动作的区别:include 动作是在页面请求访问时,将被包含页面嵌入,而 include 指令是在 JSP 页面转化成 Servlet 时才将被包含页面嵌入。

(2) <jsp:forward>

① 作用:引导请求进入新的页面。

② 基本语法为:<jsp:forward page="url" />。

(3) <jsp:plugin>

① 作用:连接客户端的 Applet 和 Bean 插件。

② 基本语法为:<jsp:plugin attribute1="value1" attribute2="value2"... >。

plugin 元素举例:

```
<jsp:plugin  type="applet" code="firstApplet" codebase="plugin/" width="640" height="
260" >
</jsp:plugin>
```

(4) <jsp:param>

① 作用:提供其他 JSP 动作的名称/值信息。

② 基本语法为:<jsp:param name="name" value="value" />。

pargm 元素举例:

```
<%@ page contentType="text/html; charset=gb2312" language="java" %>
<!DOCTYPE HTML PUBLIC "-//W3C//DTD HTML 4.0 Transitional//EN">
<HTML>
   <HEAD>
   <TITLE>param</TITLE>
   </HEAD>
   <BODY>
<%double i = Math.random();%>
   <jsp:forward page="next.jsp"><! -- 跳转到 next.jsp  -->
       <jsp:param name="number" value="<%=i%>" /><! --传递参数  -->
   </jsp:forward>
    </BODY>
</HTML>
```

(5) <jsp:useBean>

① 作用:应用 JavaBean 组件。

② 基本语法为：<jsp:useBean id="name" scope="page/request/session/application" typeSpec/>

useBean 元素举例：

useBeanTest.java 文件代码如下：

```
package xz.edu;
public class useBeanTest {
    String a = "njxz";
    public String printStr()
    {
        return a;
    }
    public String geta()
     {  return a; }
}
```

useBeanTest.jsp 文件代码如下：

```
<%@ page contentType="text/html; charset=gb2312" language="java" %>
<HTML>
  <HEAD>
  <TITLE>useBeanTest</TITLE>
  </HEAD>
  <BODY>
<jsp:useBean id="MyuseBean" class="xz.edu.useBeanTest" />
<%= MyuseBean.printStr()%>
    </BODY>
</HTML>
```

useBeanTest.jsp 文件运行效果如图 2 - 13 所示。

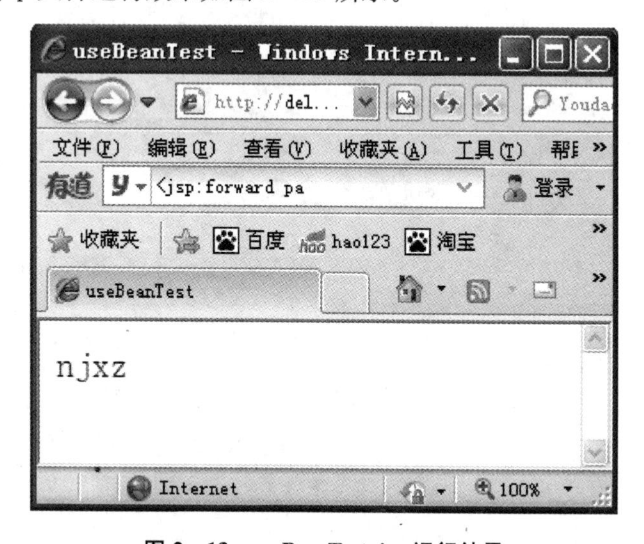

图 2 - 13　useBeanTest.jsp 运行结果

(6) <jsp:getProperty>

① 作用:将 JavaBean 的属性插入输出中。

② 基本语法为:<jsp:setProperty name="beanName" prop_expr/>

<jsp:getProperty>举例:

在上面 useBeanTest.jsp 文件中加入如下代码:

```
<jsp:getProperty name="MyuseBean" property="a"/>
```

则 useBeanTest.jsp 文件运行效果如图 2-14 所示。

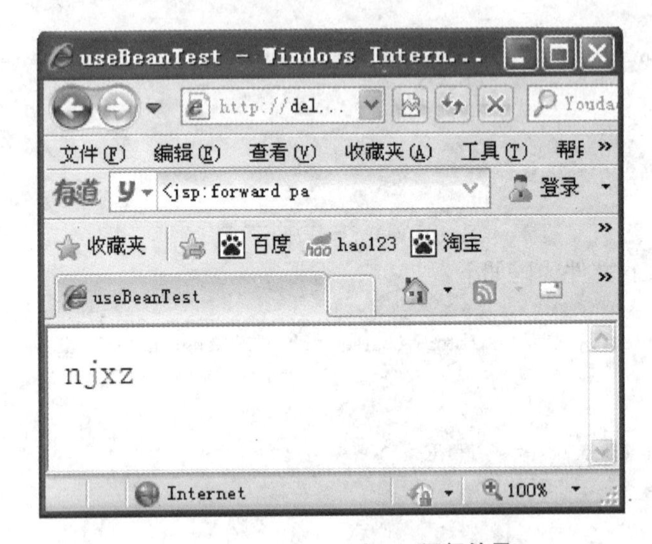

图 2-14 useBeanTest.jsp 运行结果

从上图可看到运行效果与<jsp:useBean>相同。

(7) <jsp:setProperty>

① 作用:设置 JavaBean 组件的属性值

② 基本语法为:<jsp:setProperty name="beanName" prop_expr/>

<jsp:setProperty>举例:

在上面 useBeanTest.jsp 文件中加入如下代码:

```
<jsp:setProperty name="MyuseBean" property="a" value="hello"/>
```

则 useBeanTest.jsp 文件运行效果如图 2-15 所示。

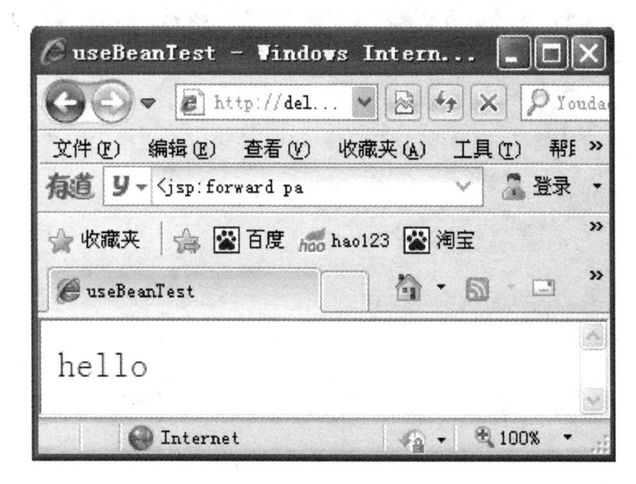

图 2-15　useBeanTest.jsp 运行结果

8. JSP 内置对象

JSP 自带了九个功能强大的内置对象，具体内容如下：

(1) request 对象

① request 对象封装了用户提交的信息。

② 通过调用该对象相应的方法可以获取封装的信息，即使用该对象可以获取用户提交信息。

③ request 对象是 HttpServletRequest 类的实例。

例：

使用 request 对象的 getParameter(string s)方法获取表单通过 text 提交的信息。如：

```
<%@ page language="java" contentType="text/html; charset=UTF-8"
     pageEncoding="UTF-8"%>
<html>
<head>
<title>request example</title>
</head>
<body>
    <form action="test.jsp">
    <input type="text" name="user">
    <input type="submit" value="Enter" name="submit">
    </form>
</body>
</html>
```

运行效果如图 2-16 所示。

图 2-16 form 表单

test.jsp 代码如下：

```
<%@ page language="java" contentType="text/html; charset=UTF-8"
    pageEncoding="UTF-8"%>
<html>
<head>
<title>request test</title>
</head>
<body>
    <p> 获取文本框提交的信息：
    <%String textContent=request.getParameter("user"); %>
    <br>
    <%=textContent %>
</body>
</html>
```

request.jsp 表单提交后，运行效果如图 2-17 所示。

图 2-17 表单提交后结果

（2）response 对象

① response 对象对客户的请求作出动态的响应，向客户端发送数据。

② response 对象是 HttpServletResponse 类的实例。

③ response 对象方法比较多，下面以重定向方法 sendRedirect (java. lang. String location)为例来简单演示如何调用该对象的方法。

response 对象举例：

responseTest. jsp 文件代码如下：

```
<%@ page language="java" contentType="text/html; charset=UTF-8"
      pageEncoding="UTF-8"%>
<html>
<head>
<title>request test</title>
</head>
<body>
<%
      String address=request. getParameter("where");
      if (address ! ="")
            response. sendRedirect("http://www. njxzc. edu. cn");
%>
      <b>
      <form action="">
            <input type="text" name="where">
            <input type="submit" value="Enter" name="submit">
      </form>
      </b>
</body>
</html>
```

该文件执行时，若文本框为空则不跳转，若不为空，则跳转到“www. njxzc. edu. cn”网页。

（3）out 对象

① out 对象是向客户端输出流进行写操作的对象。

② out 对象主要应用在脚本程序中，会通过 JSP 容器自动转换为 java. io. PrintWriter 对象。

③ out 对象具有 page 作用范围。

out 对象举例：

outTest. jsp 文件代码如下：

```
<%@ page language="java" contentType="text/html; charset=UTF-8"
    pageEncoding="UTF-8"%>
<html>
<head>
<title>request test</title>
</head>
<body>
<%
    java.util.Date now=new java.util.Date();
%>
当前时间是：
<%out.print(now); %>
</body>
</html>
```

运行结果如图 2-18 所示。

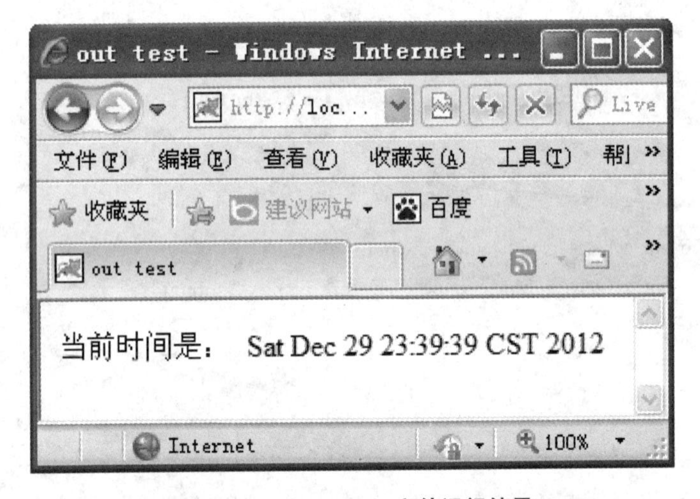

图 2-18 outTest.jsp 文件运行结果

（4）session 对象

① session 对象是"javax.servlet.httpServletSession"类的一个对象。

② session 对象提供了当前用户会话的信息和对可用于存储信息的会话范围的缓存访问，以及控制如何管理会话的方法。

③ 每个客户都对应有一个 session 对象，用来存放与这个客户相关的信息。

④ session 对象具有 session 作用范围。

session 对象有许多方法，下面以 getId 方法为例，介绍该对象的使用。

session 对象举例：

"sessionTest.jsp"文件代码如下：

```
<%@ page language="java" contentType="text/html; charset=UTF-8"
      pageEncoding="UTF-8"%>
<html>
<head>
<title>request test</title>
</head>
<body>
<%
      String s=session.getId();
%>
session 对象的 ID 是
<%out.print(s); %>
</body>
</html>
```

“sessionTest.jsp”文件运行结果如图 2-19 所示。

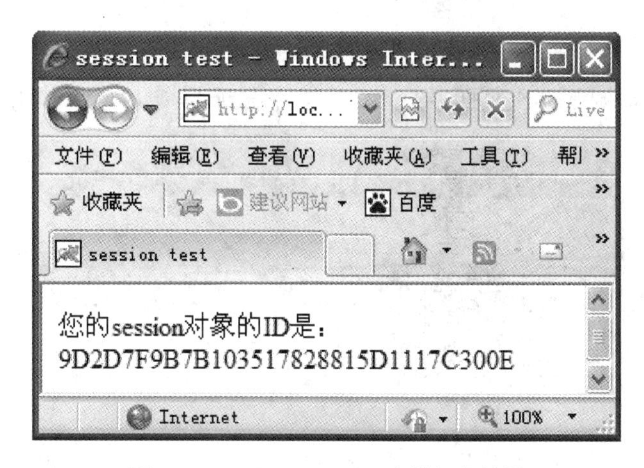

图 2-19　sessionTest.jsp 文件运行结果

(5) application 对象

① application 对象负责提供应用程序在服务器中运行时的一些全局信息。

② 当 Web 应用中的 JSP 页面开始执行时，产生一个 application 对象，所有的客户共用此对象，直到服务器关闭时才消失。

③ application 对象具有 application 作用范围。

application 举例：

“application.jsp”文件代码如下：

```
<%@ page language="java" contentType="text/html; charset=UTF-8"
    pageEncoding="UTF-8"%>
<html>
<head>
<title>application</title>
</head>
<body>
<%
    String strNum=(String)application.getAttribute("Num");
    int Num=0;
    if (strNum != null)
        Num=Integer.parseInt(strNum)+1;
    application.setAttribute("Num",String.valueOf(Num));
    out.print(Num);
%>
</body>
</html>
```

该文件执行时，每刷新一次页面，数值加 1，文件刷新 11 次后效果如图 2-20 所示。

图 2-20 application.jsp 文件刷新结果

(6) page 对象

① page 对象是 this 变量的别名，是一个包含当前 Servlet 接口引用的变量。

② page 对象具有 page 作用范围。

(7) pageContext 对象

① pageContext 对象能够存取其他内置对象，当内置对象包括属性时，也可以读取和写入这些属性。

② pageContext 是一个抽象类，实际运行的 JSP 容器必须扩展它才能被使用。

③ pageContext 对象具有 page 作用范围。

pageContext 对象举例：

pageContext 文件的代码如下：

```
<%@ page language="java" contentType="text/html; charset=UTF-8"
    pageEncoding="UTF-8"%>
<html>
<head>
<title>pageContext</title>
</head>
 <body>
<br> 使用 pageContext 设置属性 pageAttr 的值,该属性默认在 page 范围内。
<%pageContext.setAttribute("pageAttr","hello");
%>
<br>
获取属性 hello 的值:
<%=pageContext.getAttribute("pageAttr")%>
</body>
</html>
```

文件运行结果如图 2-21 所示。

图 2-21　pageContext.jsp 运行结果

(8) config 对象

① config 对象提供了对每一个服务器或者 JSP 页面的“javax.servlet.ServletConfig”对象的访问。

② config 对象中包含了初始化参数以及一些实用方法。

③ 可以为使用 web.xml 文件的服务器程序和 JSP 页面在其环境中设置初始化参数。

④ config 对象具有 page 作用范围。

下面以 config 对象的 getInit Parameter 方法为例,介绍 config 对象的使用。例:

config 对象举例:

“configTest.jsp”文件代码如下:

```
<%@ page language="java" contentType="text/html; charset=UTF-8"
    pageEncoding="UTF-8"%>
<html>
<head>
<title>configTest</title>
</head>
<body>
<br>
Photo:
<%=config.getInitParameter("photo")%>
</body>
</html>
```

web.xml 文件代码如下：

```
<?xml version="1.0" encoding="UTF-8"?>
<web-app version="2.5"
    xmlns="http://java.sun.com/xml/ns/javaee"
    xmlns:xsi="http://www.w3.org/2001/XMLSchema-instance"
    xsi:schemaLocation="http://java.sun.com/xml/ns/j2ee
    http://java.sun.com/xml/ns/javaee/web-app_2_5.xsd">
  <servlet>
    <servlet-name>admin</servlet-name>
    <jsp-file>/configTest.jsp</jsp-file>
    <init-param>
        <param-name>photo</param-name>
        <param-value>88888888</param-value>
    </init-param>
  </servlet>
    <servlet-mapping>
    <servlet-name>admin</servlet-name>
    <url-pattern>/configTest.jsp</url-pattern>
  </servlet-mapping>
</web-app>
```

“configTest.jsp”文件执行效果如图 2－22 所示。

图 2-22　configTest.jsp 运行结果

(9) exception 对象

① exception 对象是异常对象。与错误不同，这里的异常指的是 Web 应用程序中所能够识别并处理的问题。如果在 JSP 页面中没有捕捉到异常，就会产生 exception 对象，并把这个对象传递到在 page 设定的 errorpage 中去，然后在 errorpage 页面中处理相应的 exception。

② exception 对象具有 page 作用范围。

③ 需要注意的是：要使用内置的 exception 对象，必须在 page 命令中设定<%@page isErrorPage="true" %>，否则会出现编译错误。

exception 对象举例：

"error.jsp"文件代码如下：

```
<%@ page language="java" contentType="text/html; charset=UTF-8"
    pageEncoding="UTF-8" isErrorPage="true" %>
<html>
<head>
<title>error</title>
</head>
<body>
<br>
抱歉，程序发送异常！
</body>
</html>
```

还需在"web.xml"文件中增加如下代码：

```
<error-page>
        <error-code>500</error-code>
        <location>/error.jsp</location>
  </error-page>
```

“errorTest.jsp”文件代码如下：

```
<%@ page language="java" contentType="text/html; charset=UTF-8"
    pageEncoding="UTF-8"  %>
<!DOCTYPE html PUBLIC "-//W3C//DTD HTML 4.01 Transitional//EN" "http://www.w3.
org/TR/html4/loose.dtd">
<html>
<head>
<title>error Test</title>
</head>
<body>
当前时间是：
<%out.print(now); %>
</body>
</html>
```

因为该文件中 now 对象为空，因此该文件运行后将跳转到 error.jsp 文件，运行结果如图 2-23 所示。

图 2-23 errorTest.jsp 文件运行结果

9. Java Bean

JavaBean 是 Java 的一种软件组件模型，是 Sun Microsystems 公司为了适应网络计算提出的。按照属性作用的不同，一般可以分为四类：

(1) 简单属性：表示一个伴随 get 和 set 方法的变量。属性的名称与该属性相关的 get、set 方法对应。

(2) 索引属性：表示一个数组值。使用与该属性对象的 set 和 get 方法可以存取数组中某个元素的数值。

(3) 绑定属性：当绑定属性发生变化时，必须通知其他 JavaBean 组件对象。

(4) 约束属性：约束属性值的变化首先要被所有的监听器验证之后，才有可能真正发生改变。

JSP 提供了三种标记来使用 JavaBean：

① 初始化 Bean

使用标记<jsp:useBean/>

② 获取 Bean 属性

使用标记<jsp:getProperty/>

③ 设置 Bean 属性

使用标记<jsp:setProperty/>

JavaBean 的生命周期分为四种范围:page、request、session 和 application,通过设置 JavaBean 的 scope 属性,可以对 JavaBean 设置不同的生命周期,它们覆盖的范围如图 2-24 所示。

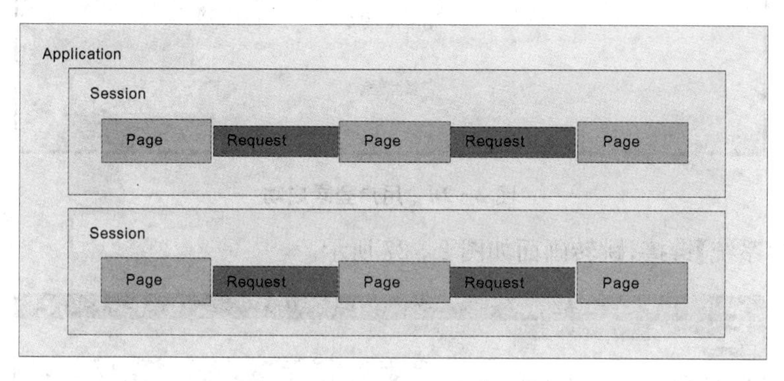

图 2-24 JavaBean 生命周期

§2.4 JSP 应用实例

2.4.1 应用简介

本实例是用 JSP、JavaBean 开发一个小型在线用户统计系统,View 页面用于信息的录入、显示,JavaBean 通过 Session 对象计算当前在线用户信息。

首先,用户登录界面如图 2-25 所示。

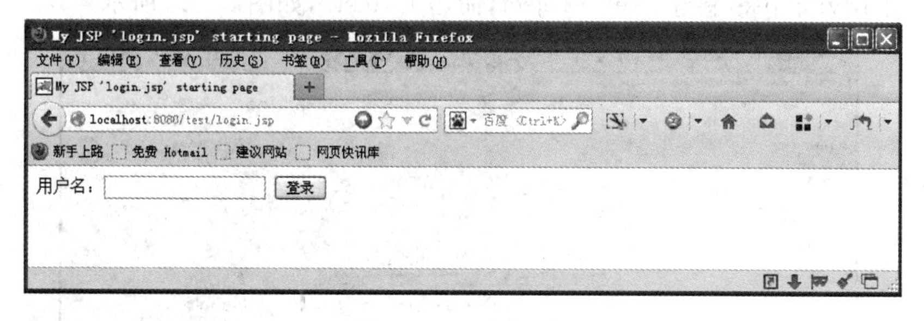

图 2-25 用户登录

登录成功后显示当前在线人数,如图 2-26 所示。

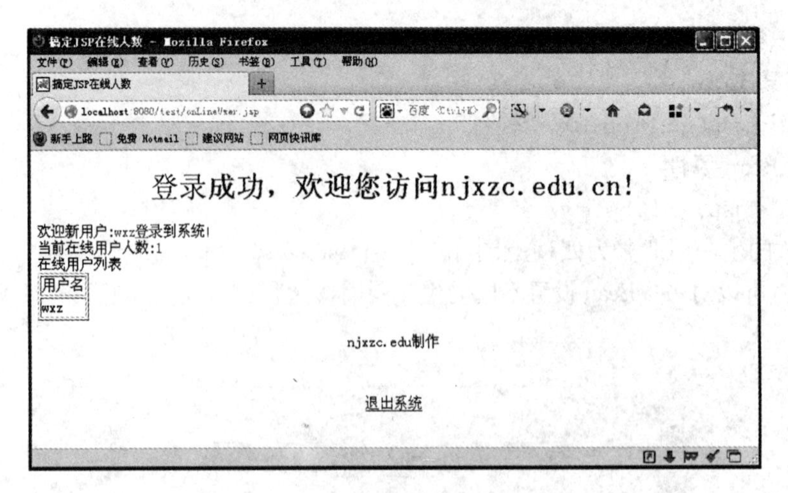

图 2-26　用户登录成功

单击【退出系统】链接，跳转画面如图 2-27 所示。

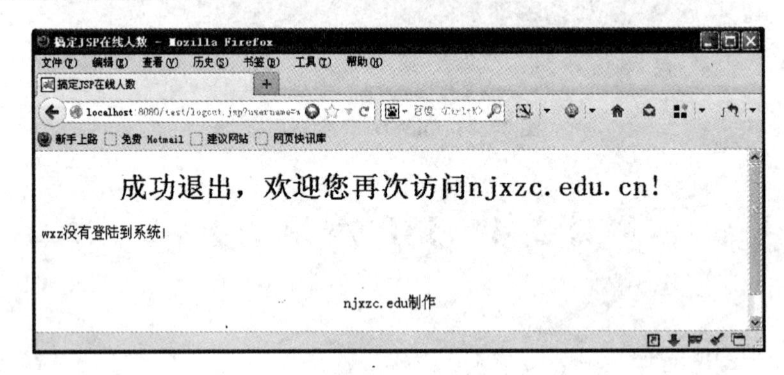

图 2-27　用户退出

2.4.2　项目创建步骤

(1) 打开 MyEclipse，新建 Web Project，命名为"test"，如图 2-28 所示。

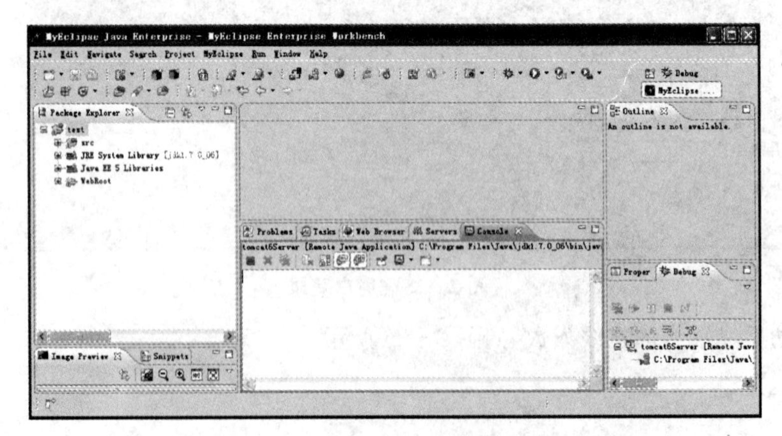

图 2-28　新建 Web Project

(2) 建立 JavaBean。

因为要统计在线人数，需要使用 Session 对象，如果一个 JavaBean 实现了 HttpSessionBindingListener

接口,当这个对象被绑定到 Session 中或者从 Session 中被删除时,Servlet 容器会通知这个对象,而这个对象在接收到通知后,可以做一些初始化或清除状态的操作。该 JavaBean 代码如下:

```
package xz;
import javax.servlet.http.*;
import javax.servlet.*;
import java.util.*;
public class onLineUser implements HttpSessionBindingListener {
    public onLineUser(){
    }
    private Vector users=new Vector();
    public int getCount(){
        users.trimToSize();
        return users.capacity();
    }
    public boolean existUser(String userName){
        users.trimToSize();
        boolean existUser=false;
        for (int i=0;i<users.capacity();i++ )
        {
            if (userName.equals((String)users.get(i)))
            {
                existUser=true;
                break;
            }
        }
        return existUser;
    }
    public boolean deleteUser(String userName) {
        users.trimToSize();
        if(existUser(userName)){
            int currUserIndex=-1;
            for(int i=0;i<users.capacity();i++){
                if(userName.equals((String)users.get(i))){
                    currUserIndex=i;
                    break;
                }
            }
            if (currUserIndex! =-1){
                users.remove(currUserIndex);
                users.trimToSize();
                return true;
```

```
        }
      }
      return false;
    }
    public Vector getOnLineUser()
    {
      return users;
    }
    public void valueBound(HttpSessionBindingEvent e) {
      users.trimToSize();
      if(!existUser(e.getName())){
        users.add(e.getName());
        System.out.print(e.getName()+"\t 登录系统\t"+(new Date()));
        System.out.println(" 在线用户数为:"+getCount());
      }else
        System.out.println(e.getName()+"已经存在");
    }
    public void valueUnbound(HttpSessionBindingEvent e) {
      users.trimToSize();
      String userName=e.getName();
      deleteUser(userName);
      System.out.print(userName+"\t 退出系统\t"+(new Date()));
      System.out.println(" 在线用户数为:"+getCount());
    }
}
```

(3) 创建登录界面,代码如下:

```
<%@ page language="java" import="java.util.*" pageEncoding="GB18030"%>
<%
String path = request.getContextPath();
String basePath = request.getScheme()+"://"+request.getServerName()+":"+request.
getServerPort()+path+"/";
%>

<!DOCTYPE HTML PUBLIC "-//W3C//DTD HTML 4.01 Transitional//EN">
<html>
  <head>
    <base href="<%=basePath%>">
    <title>My JSP "login.jsp" starting page</title>
  </head>
  <body>
```

```
        <form action="onLineUser.jsp" method="post">
            用户名:<input type="text" name="username"/>
                <input type="submit" value="登录"/>
        </form>
    </body>
</html>
```

(4) 创建显示人数界面"onLineUser.jsp"文件,代码如下:

```
<%@ page language="java" import="java.util.*" pageEncoding="GB18030"%>
<%@ page import="xz.onLineUser" %>
<jsp:useBean id="onlineuser" class="xz.onLineUser" scope="application"/>
<html>
    <head>
        <title>在线人数</title>
    </head>
    <body>
        <center>
            <p><h1>登录成功,欢迎您访问 njxzc.edu.cn! </h1></p>
        </center>
        <% session = request.getSession(false); %>
        <%
            String username=request.getParameter("username");
            if (onlineuser.existUser(username)){
                out.println("用户<font color=red>"+username+"</font>已经登录!");
            }else{
                session.setMaxInactiveInterval(50); file://Sesion 有效时长,以秒为单位
                session.setAttribute(username,onlineuser);
                out.println("欢迎新用户:<font color=red>"+username+"</font>登录
录到系统!");
            }
            out.println("<br>当前在线用户人数:<font color=red>"+onlineuser.getCount
()+"</font><br>");
            Vector vt=onlineuser.getOnLineUser();
            Enumeration e = vt.elements();
            out.println("在线用户列表");
            out.println("<table border=1>");
            out.println("<tr><td>用户名</td></tr>");
            while(e.hasMoreElements()){
                out.println("<tr><td>");
                out.println((String)e.nextElement()+"<br>");
                out.println("</td></tr>");
            }
```

```
                out.println("</table>");
            %>
            <center>
                <p>njxzc.edu 制作</p>
            <p> </p>
            <%
                out.println("<p><a href=logout.jsp? username="+username+">退出系统
</a></p>");
            %>
            </center>
        </body>
</html>
```

(5) 创建登录页面"Logout.jsp"文件,代码如下:

```
<%@ page language = "java" import = "xz.onLineUser, java.util.*" pageEncoding = "
GB18030"%>
<%@ page contentType="text/html;charset=gb2312" %>
<jsp:useBean id="onlineuser" class="xz.onLineUser" scope="application"/>
<html>
    <head>
        <title>搞定 JSP 在线人数</title>
    </head>
    <body>
        <center>
            <p><h1>成功退出,欢迎您再次访问 njxzc.edu.cn! </h1></p>
        </center>
        <%
            String username=request.getParameter("username");
            if(onlineuser.deleteUser(username))
            out.println(username+"已经退出系统!");
            else
                out.println(username+"没有登录到系统!");
        %>
        <center>
            <p>njxzc.edu 制作</p>
            <p> </p>
        </center>
    </body>
</html>
```

(6) 部署和运行

部署 test 项目到 Tomcat 的 webapps 目录下,启动 Tomcat 之后,打开浏览器,在地址栏输入网址"http://localhost:8080/test/login.jsp",然后回车,浏览器出现如图 2-25 所示界面,即表示运行成功。

巩固练习

1. 画出 HTML 文件的基本框架图。
2. 写出 JSP 的内置对象及作用。
3. 进一步了解 JavaBean 和 JSP 的关系。
4. 深入了解 JSP 的工作原理。

第3章　Struts 2 概述及基本应用

学习目标

1. 了解 MVC 的核心思想。
2. 了解 Struts 2 工作流程。
3. 了解 Struts 2 中的基本元素作用。
4. 掌握 Struts 2 的数据验证方法。
5. 掌握 Struts 2 标签库的组成。
6. 掌握 Struts 2 拦截器的作用。
7. 掌握 Struts 2 文件上传功能。

Struts 2 是 Struts 1 的下一代产品，是在 Struts 1 和 WebWork 两个 MVC 框架基础上进行了合并的全新的 Struts 2 框架。其全新的体系结构与 Struts 1 的体系结构差别甚大。Struts 2 以 WebWork 为核心，采用拦截器的机制来处理用户的请求，这样的设计也使得业务逻辑控制器能够与 Servlet API 完全脱离开，所以 Struts 2 可以理解为 WebWork 的更新产品。虽然从 Struts 1 到 Struts 2 有着太大的变化，但是相对于 WebWork，Struts 2 的变化很小。因为 Struts 2 是 WebWork 的升级，而不是一个全新的框架，因此稳定性、性能等各方面都有很好的保证，而且吸收了 Struts 1 和 WebWork 两者的优势。因此，它是一个值得学习和了解的框架。

§3.1　Struts 2 概述

3.1.1　MVC 思想概述

1. MVC 基本介绍

MVC 架构的核心思想是：将程序分成相对独立，而又能协同工作的三个部分。通过使用 MVC 架构，可以降低模块之间的耦合，提供应用的可扩展性。另外，MVC 的每个组件只关心组件内的逻辑，不应与其他组件的逻辑混合。MVC 并不是 Java 所独有的概念，而是面向对象程序都应该遵守的设计理念。

2. JSP Model 1 和 Model 2 介绍

(1) JSP 设计模式 1

在 JSP 技术的发展初期，由于其便于快速开发的优点，很快就成为了创建 Web 的应用的热门技术之一。在 JSP 页面中可以很容易实现内容的显示。业务逻辑的编写以及流程的控制，从而快速地完成应用开发。最初很多的 Java Web 应用甚至全部由 JSP 页面构成，这种以

JSP 为中心的开发模型称为 Model 1（设计模式 1）。具体实现方式如图 3-1 所示。

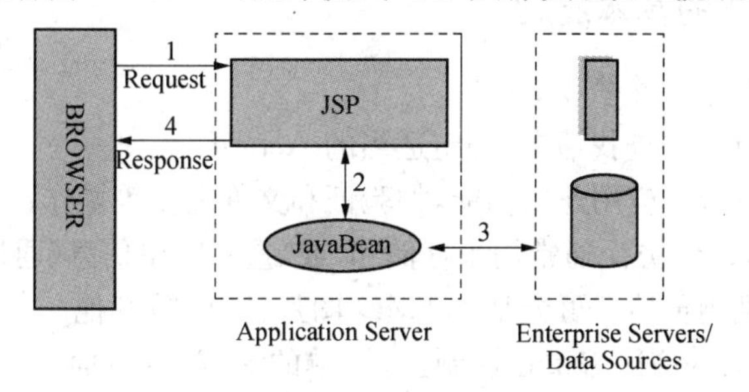

图 3-1　Model 1

在设计模式 1 中，JSP 页面负责接收处理客户端的 Web 浏览器发送的请求，并在处理后直接进行响应。期间，一般会借助 Java Bean 处理复杂的业务逻辑。

(2) JSP 设计模式 2

Model 2 是 MVC 设计模式的一种具体实现方式。在 Model 2 中采用 Servlet 作为控制器，负责接收客户端 Web 浏览器发送的所有请求，并依据处理的不同结果，转发到对应的 JSP 页面实现在浏览器客户端显示。具体实现方式如图 3-2 所示。

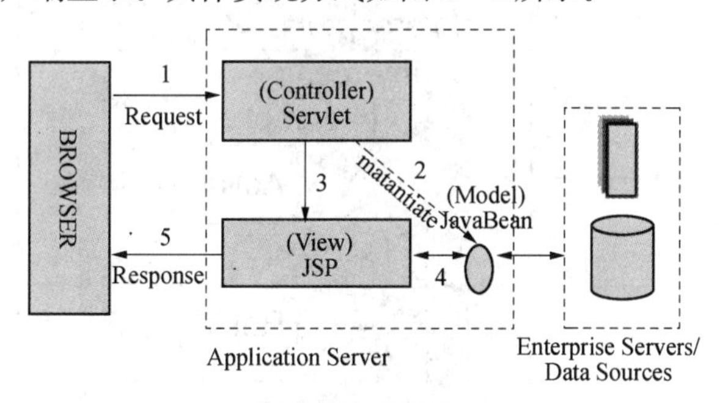

图 3-2　Model 2

MVC(Model-View-Controller)模式，即模型-视图-控制器模式，其核心思想是将整个程序代码分成相对独立而又能协同工作的三个组成部分，这三个部分分别是：

① 模型(Model)：业务逻辑层。实现具体的业务逻辑、状态管理的功能。

② 视图(View)：表示层。即与用户实现交互的界面，通常实现数据的输入和输出功能。

③ 控制器(Controller)：控制层。起到控制整个业务流程(Flow Control)的作用，实现 View 和 Model 部分的协同工作。

MVC 设计模式的结构及各组成部分间的通信方式如图 3-3 所示。

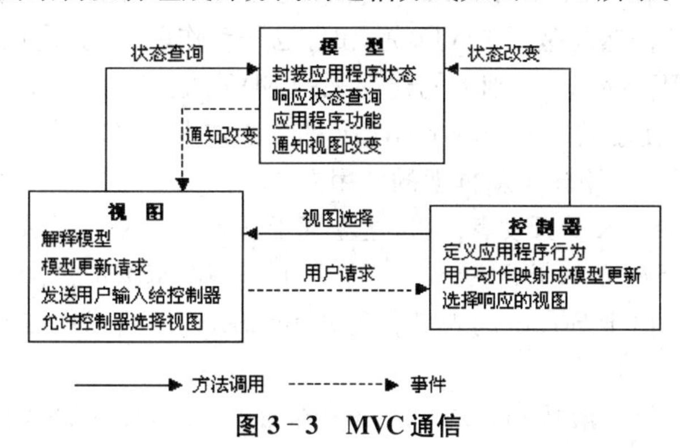

图 3-3　MVC 通信

3.1.2　Struts 2 的基本流程

(1) Web 浏览器请求一个资源；

(2) 过滤器 Dispatcher 查找请求，确定适当的 Action；

(3) 拦截器自动对请求应用通用功能，如验证和文件上传等操作；

(4) Action 的 execute 方法通常用来存储和(或)重新获得信息(通过数据库)；

(5) 结果被返回到浏览器。可能是 HTML、图片、PDF 或其他。

其实，Struts 2 框架的应用着重在控制上，简单的流程是：页面→控制器→页面。最重要的是控制器的取数据与处理后传数据的问题。Struts 2 的体系结构参考图 3-4，更直观地展现出其流程。

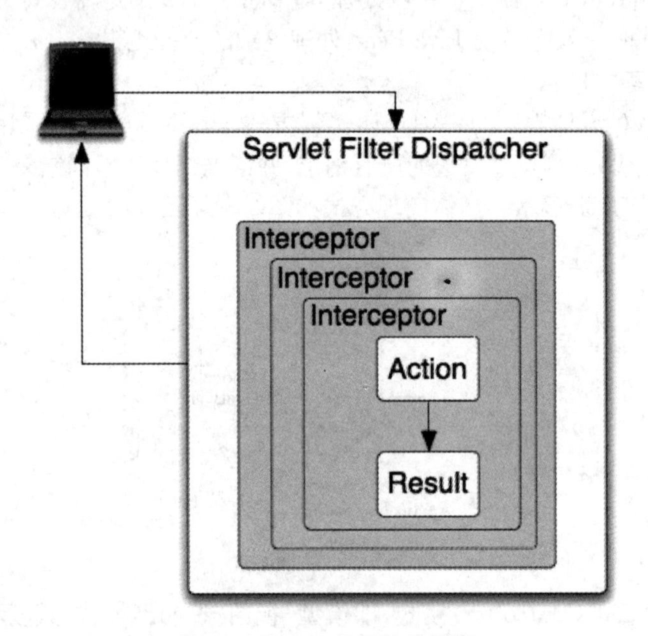

图 3-4　Struts 2 的体系结构

§3.2　Struts 2 简单实例开发及工作流程

Struts 2 虽然仍可看作 Struts 的第二个版本，但在配置和使用上已经与 Struts 1.x 相差太大。(当然，Struts 2 仍然是基于 MVC 模式的，也是动作驱动的，这也是唯一没变的地方) Struts 2 实际上是在 WebWork 基础上构建起来的 MVC 框架。从 Struts 2 的源代码中可以看到，有很多都是直接使用的 xwork(WebWork 的核心技术)的包。从技术上来说 Struts 2 是全新的框架，接下来学习一下这个新框架的使用方法。

对于使用过 Struts 1.x 的开发者而言，建立基于 Struts 1.x 的 Web 程序的基本步骤非常清楚。接下来回顾一下建立基于 Struts 1.x 的 Web 程序的基本步骤：

(1) 安装 Struts。由于 Struts 的入口点是 ActionServlet，所以得在 web.xml 文件中配置一下这个 Servlet。

(2) 编写 Action 类(一般从 org.apache.struts.action.Action 类继承)。

(3) 编写 ActionForm 类(一般从 org. apache. struts. action. ActionForm 类继承),这一步不是必须的,如果要接收客户端提交的数据,需要执行这一步。

(4) 在 struts—config. xml 文件中配置 Action 和 ActionForm。

(5) 如果要采集用户输入的数据,一般需要编写若干 JSP 页面,并通过这些 JSP 页面中的 form 将数据提交给 Action。

接下来编写一个基于 Struts 2 的 Web 应用实例,在创建该应用实例过程中,会把每个步骤与创建 Struts 1 的应用实例步骤做适当比较。这个应用实例的功能是让用户输入两个整数,并提交给一个 Struts Action,并计算这两个数的代数和,如果代码和为非负数,则跳转到 positive. jsp 页面,否则跳转到 negative. jsp 页面。

1. 下载 Struts 2

登录"http://struts. apache. org/",下载 struts - 2. x. x-all. zip。下载完后解压文件,开发 struts 2 应用需要依赖的". jar"文件在解压目录的 lib 文件夹下,不同的应用需要的 JAR 包是不同的。下面给出了开发 Struts 2 程序最少需要的 JAR,分别是:

① struts2-core-2. x. x. jar :Struts 2 框架的核心类库;

② xwork-core-2. x. x. jar :XWork 类库,Struts 2 使用 WebWork 的核心技术的类库;

③ ognl-3. 0. x. jar :对象图导航语言(Object Graph Navigation Language),Struts 2 框架使用的一种表达式语言;

④ freemarker-2. 3. x. jar :Struts 2 的 UI 标签的模板使用 FreeMarker 编写;

⑤ commons-logging-1. 1. x. jar :ASF 出品的日志包,Struts 2 框架使用这个日志包来支持 Log4J 和 JDK 1. 4+的日志记录。

随着版本的不同,所需包名和数目都有所差异,因此为避免出现 JAR 包的现象,最简单地做法是:

如果不需要跟第三方框架集成,把不带-plugin 结尾的". jar"文件都添加入类路径即可;如果需要跟第三方框架集成,这时候还需要加入对应的-plugin 结尾的". jar"文件。例如跟 spring 集成,需要加入 struts 2-spring-plugin-2. x. x. jar.

2. 建立项目并配置 MyEclipse

打开 MyEclipse,建立一个 Web 项目,命名为"FirstStruts 2"。加载 Struts 2 基本类库,下面把这几个类库添加到项目中。

右击项目名,选择【Build Path】→【Configure Build Path】菜单项,出现如图 3 - 5 所示的对话框。单击【Add External JARs】按钮,进入下载的 Struts 2 目录的 lib 文件夹,选中包括上面的五个在内的七个 Jar 包,单击【OK】按钮完成类库的添加。

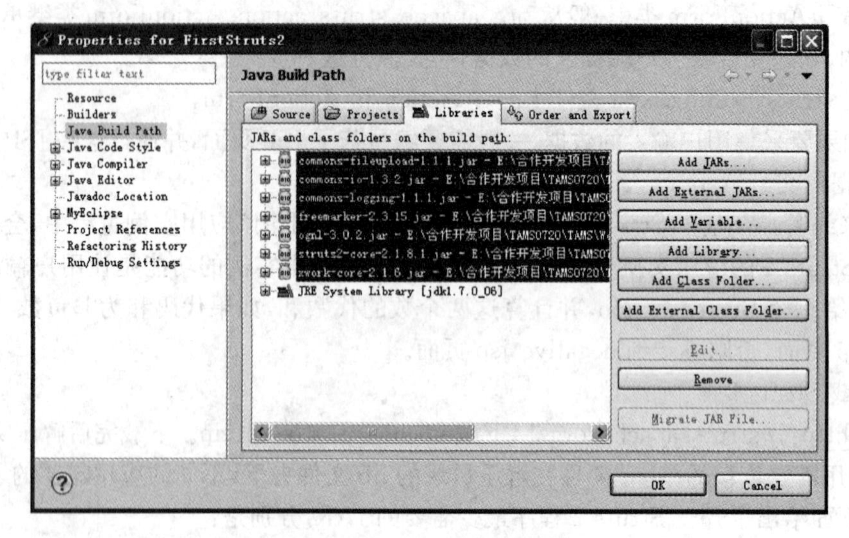

图 3-5 MyEclipse 添加类库

3. 配置 web. xml

这一步对于 Struts 1 和 Struts 2 都是必须的，只是安装的方法不同。Struts 1 的入口点是一个 Servlet，而 Struts 2 的入口点是一个过滤器(Filter)。因此，Struts 2 要按过滤器的方式配置，web. xml 文件代码如下：

```
<?xml version="1.0" encoding="UTF-8"?>
<web-app version="2.5"
    xmlns="http://java.sun.com/xml/ns/javaee"
    xmlns:xsi="http://www.w3.org/2001/XMLSchema-instance"
    xsi:schemaLocation="http://java.sun.com/xml/ns/javaee
    http://java.sun.com/xml/ns/javaee/web-app_2_5.xsd">
  <welcome-file-list>
    <welcome-file>index.jsp</welcome-file>
  </welcome-file-list>
  <filter>
    <filter-name>struts 2</filter-name>
    <filter-class>
        org.apache.struts2.dispatcher.FilterDispatcher
    </filter-class>
  </filter>
  <filter-mapping>
    <filter-name>struts 2</filter-name>
    <url-pattern>/*</url-pattern>
  </filter-mapping>
</web-app>
```

4. 编写 Action 类

这一步在 Struts 1. x 中也必须进行。只是 Struts 1. x 中的动作类必须从 Action 类中继承，而 Struts 2. x 的动作类需要从 com. opensymphony. xwork2. ActionSupport 类继承。下面是计算两个整数代码和的 Action 类，代码如下：

```
package action;
import com.opensymphony.xwork2.ActionSupport;
public class FirstAction extends ActionSupport{
    private static final long serialVersionUID = 1L;
    private int operand1;
    private int operand2;
    private int sum;
    public String execute() throws Exception{
        if(getSum()>=0) //如果代码数和是非负整数,跳到 positive.jsp 页面
        {
            return"positive";
        }
        else //如果代码数和是负整数,跳到 negative.jsp 页面
        {
            return"negative";
        }
    }
    public int getOperand1()
    {
        return operand1;
    }
    public void setOperand1(int operand1)
    {
        System.out.println(operand1);
        this.operand1=operand1;
    }
    public int getOperand2()
    {
        return operand2;
    }
    public void setOperand2(int operand2)
    {
        System.out.println(operand2);
        this.operand2=operand2;
    }
    public int getSum()
    {
        return operand1+operand2; //计算两个整数的代码数和
    }
}
```

从上面的代码可以看出，动作类的一个特征就是要重写 execute 方法，只是 Struts 2 的 execute 方法没有参数了，而 Struts 1. x 的 execute 方法有四个参数。而且 execute 方法的返回值也不同的。Struts 2 只返回一个 String，用于表述执行结果（就是一个标志）。在 Struts 1. x 中，必须要单独建立一个 ActionForm 类（或是定义一个动作 Form），而在 Struts 2 中 ActionForm 和 Action 已经合二为一了。从步骤的代码可以看出，后面的部分就是应该写在 ActionForm 类中的内容，所以本例的 ActionForm 类已经编写完成（就是 Action 类的后半部分）。

5. 创建并配置 Struts. xml

这一步 Struts 1. x 和 Struts 2. x 都是必需的，只是在 Struts 1. x 中的配置文件一般叫 Struts-config. xml（当然也可以是其他的文件名），而且一般放到 WEB－INF 目录中。而在 Struts 2. x 中的配置文件一般为 Struts. xml，放到 src 目录中。下面是在 Struts. xml 中配置动作类的代码：

```
<?xml version="1.0" encoding="UTF-8"?>
<!DOCTYPE struts PUBLIC
    "-//Apache Software Foundation//DTD Struts Configuration 2.0//EN"
    "http://struts.apache.org/dtds/struts-2.0.dtd">
    <struts>
    <package name="struts 2" namespace="/mystruts" extends="struts-default">
        <action name="sum" class="action.FirstAction">
            <result name="positive">/positive.jsp</result>
            <result name="negative">/negative.jsp</result>
        </action>
    </package>
</struts>
```

在<struts>标签中可以有多个<package>，第一个<package>可以指定一个 Servlet 访问路径（不包括动作名），如“/FirstStruts 2”。extends 属性继承一个默认的配置文件“struts-default”，一般都继承于它，<action>标签中的 name 属性表示动作名，class 表示动作类名。<result>标签的 name 实际上就是 execute 方法返回的字符串，如果返回的是“positive”，就跳转到 positive. jsp 页面，如果是“negative”，就跳转到 negative. jsp 页面。在<struts>中可以有多个<package>，在<package>中可以有多个<action>。可以用如下的 URL 来访问这个动作：http://localhost:8080/FirstStruts 2/sum. action。

注：Struts 1. x 的动作一般都以". do"结尾，而 Struts 2 是以". action"结尾。

6. 编写用户录入接口（JSP 页面）

在 Web 根目录建立一个 sum. jsp，代码如下：

```
<%@ page language="java" import="java.util.*" pageEncoding="GB2312"%>
<%@taglib prefix="s" uri="/struts-tags"%>
<!DOCTYPE HTML PUBLIC "-//W3C//DTD HTML 4.01 Transitional//EN">
<html>
    <head>
```

```
        <title>输入操作数</title>
    </head>
    <body>
        求代数和  <br/>
      <s:form action="mystruts/sum.action">
            <s:textfield name="operand1" label="操作数 1"/>
            <s:textfield name="operand2" label="操作数 2"/>
                <s:submit value="代数和"/>
      </s:form>
    </body>
</html>
```

在 sum.jsp 中使用了 Struts 2 带的标签库在 Struts 2 中已经将 Struts 1.x 的好几个标签库都统一了，在 Struts 2 中只有一个标签库/struts—tags.这里面包含了所有的 Struts 2 标签。但使用 Struts 2 的标签要注意一下。在<s:form>中最好都使用 Struts 2 标签，尽量不要用 HTML 或普通文本的标签。

接着，创建 positive.jsp 文件，代码如下：

```
<%@ page language="java" import="java.util.*" pageEncoding="GB2312"%>
<%@taglib prefix="s" uri="/struts-tags"%>
<!DOCTYPE HTML PUBLIC "-//W3C//DTD HTML 4.01 Transitional//EN">
<html>
    <head>
        <title>显示代数和</title>
    </head>
    <body>
        代数和为非负整数  <br/>
       <h1><s:property value="sum"/></h1>
    </body>
</html>
```

最后，创建 negative.jsp 文件，代码如下：

```
<%@ page language="java" import="java.util.*" pageEncoding="GB2312"%>
<%@taglib prefix="s" uri="/struts-tags"%>
<!DOCTYPE HTML PUBLIC "-//W3C//DTD HTML 4.01 Transitional//EN">
<html>
    <head>
        <title>显示代数和</title>
    </head>
    <body>
        代数和为负整数  <br/>
       <h1><s:property value="sum"/></h1>
```

```
    </body>
</html>
```

这两个 jsp 页面的实现代码基本一样，只使用了一个<s:property>标签来显示 Action 类中的 sum 属性值。<s:property>标签是从 request 对象中获得了一个对象中得到的 sum 属性。

7. 部署运行

部署 FirstStuts 2 项目到 Tomcat 的 webapps 目录下，启动 Tomcat 之后，打开浏览器，在地址栏输入“http://localhost:8080/ FirstStuts 2/sum.jsp”，然后回车，浏览器显示结果如图 3-6 所示。

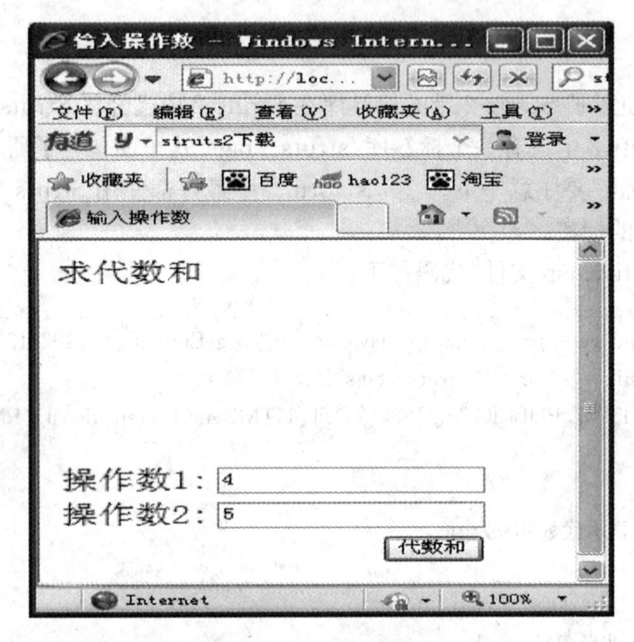

图 3-6　sum.jsp 页面

如果输入两个数和为负数，则显示结果如图 3-7 所示。

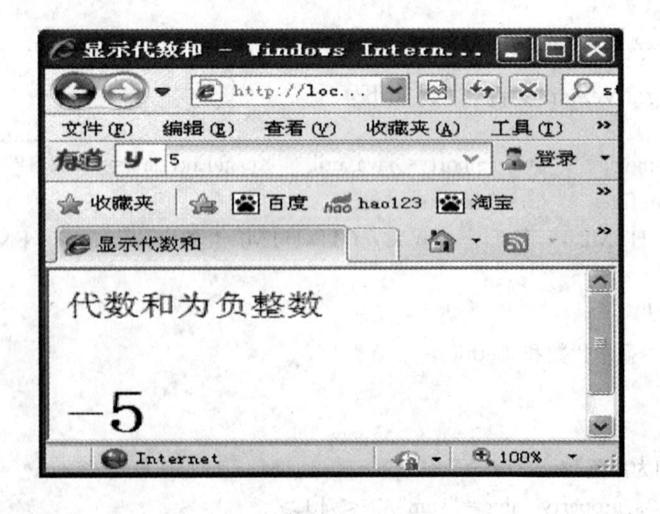

图 3-7　负数显示

如果输入两个数和为正数，则显示结果如图 3-8 所示，即表示运行成功。

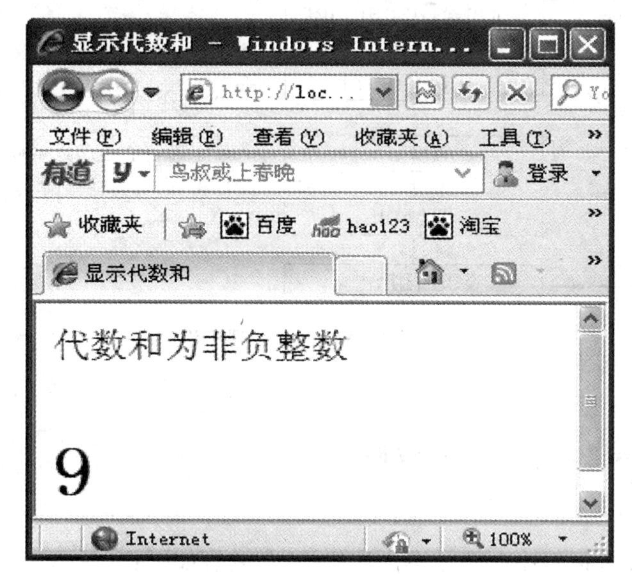

图 3-8　正数显示

§3.3　Struts 2 的工作流程及文件详解

3.3.1　Struts 2 的工作流程

当用户发送一个请求后，也就是一个"*.action"，"web.xml"文件中的配置 FilterDispatcher 就会过滤该请求。如果请求时以".acton"结尾，该请求就会被转入 Struts 2 框架处理。Struts 2 框架中的配置文件"struts.xml"会起映射作用，它会根据"*"来决定调用用户定义的哪个 Action 类。

例如在项目 FirstStruts 2 中，请求为 sum.action，前面"*"的部分是"sum"，所以在"struts.xml"中有个 Action 类的 name 为"sum"，这表示该请求与这个 Action 匹配，就会调用该 Action 中 class 属性指定的 Action 类。但是在 Struts 2 中，用户定义的 Action 类并不是业务控制器，而是 Action 代理，其并没有和 Servlet API 耦合。所以 Struts 2 框架提供了一系列的拦截器，它负责将 HttpServletRequest 请求中的请求参数解析出来，传入用户定义的 Action 类中。然后再调用 execute()方法处理用户请求，处理结束后，会返回一个值。这时，Struts 2 框架的"struts.xml"文件又起到映射作用，会根据其返回值决定跳转到哪个页面。整个处理流程如图 3-9 所示。

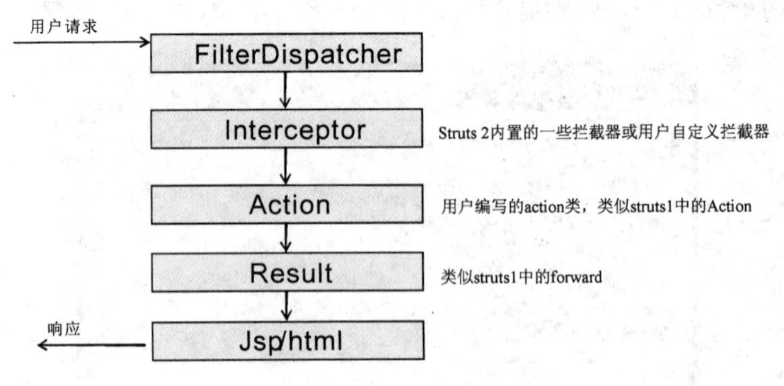

图 3-9 Struts 2 处理流程

3.3.2 Struts 2 中各种文件详解

1. web.xml 文件

文件中的 webapp 标签中配置了下面这样一段，内容如下：

```
...
<filter>
    <filter-name>struts 2</filter-name>
    <filter-class>org.apache.struts 2.dispatcher.FilterDispatcher</filter-class>
</filter>
<filter-mapping>
    <filter-name>struts 2</filter-name>
    <url-pattern>/*</url-pattern>
</filter-mapping>
...
```

可以看出，上述代码中配置了一个过滤器，接下来介绍一下过滤器的使用。

Filter 过滤器是实现了 java.Serlvet.Filter 接口的应用程序，过滤器可以自定义，也可以是框架自带的，它运行在用户请求和处理程序之间。过滤器会对用户请求和处理程序响应的内容进行处理，主要用于权限控制、编码转换等。

所有过滤器必须实现 java.Serlvet.Filter 接口，该接口中含有三个过滤器类必须实现的方法：

(1) init(FilterConfig)：Servlet 过滤器的初始化方法，Servlet 容器创建 Servlet 过滤器实例后将调用这个方法。

(2) doFilter(ServletRequest,ServletResponse,FilterChain)：完成实际的过滤操作，当用户请求与过滤器关联的 URL 时，Servlet 容器将先调用过滤器的 doFilter 方法，返回响应之前也会调用此方法。FilterChain 参数用于访问过滤器链上的下一个过滤器。

(3) destroy()：Servlet 容器在销毁过滤器实例前调用该方法，这个方法可以释放 Servlet 过滤器占用的资源，过滤器类实现后，需在"web.xml"文件中进行配置，内容如下：

```
<filter>
    <!--自定义的名称-->
    <filter-name>过滤器名</filter-name>
    <!--自定义的过滤器类,注意:这里要在包下,过滤器的类名前要加包名-->
    <filter-class>过滤器对应类</filter-class>
    <init-param>
        <! --类中参数名称-->
        <param-name>参数名称</param-name>
        <! --对应参数的值-->
        <param-value>参数值</param-value>
    </init－param>
</filter>
```

过滤器的关联方式有三种:关联一个 URL 资源、关联一个 URL 目录下的所有资源、关联一个 Servlet。

① 关联一个 URL 资源:

```
<filter-mapping>
    <!--这里与上面配置的名称要相同-->
    <filter-name>过滤器名</filter-name>
    <!--与该 URL 资源关联-->
    <url-pattern>xxx. jsp</url-pattern>
</filter-mapping>
```

② 关联一个 URL 目录下的所有资源:

```
<filter-mapping>
    <filter-name>过滤器名</filter-name>
    <url-pattern>/ * </url-pattern>
</filter-mapping>
```

③ 关联一个 Servlet:

```
<filter-mapping>
    <filter-name>过滤器名</filter-name>
    <url-pattern>Servlet 名称</url-pattern>
</filter-mapping>
```

2. struts. xml 文件

struts. xml 文件通常放在 Web 应用程序的 WEB－INF/classes 目录下,该目录下的 struts. xml 文件将被 Struts 2 框架自动加载。

struts. xml 文件是一个 XML 文件,文件前面是 XML 的头文件,然后是<struts>标签,位于 Struts 2 配置的最外层,其他标签都是包含在里面的。

在大部分应用里,随着应用规模的增加,系统中 Action 数量也大量增加,导致 struts. xml

配置文件变得非常臃肿。为了避免 struts. xml 文件过于庞大、臃肿，提高 struts. xml 文件的可读性，可以将一个 struts. xml 配置文件分解成多个配置文件，然后在 struts. xml 文件中包含这些配置文件。下面的 struts. xml 通过<include>元素包含指定的多个配置文件：

```
<?xml version="1.0" encoding="UTF-8"?>
<!DOCTYPE struts PUBLIC
      "-//Apache Software Foundation//DTD Struts Configuration 2.0//EN"
      "http://struts.apache.org/dtds/struts-2.0.dtd">
<struts>
      <include file="struts-login.xml"/>
      <include file="struts-operate.xml"/>
</struts>
```

通过这种方式，就可将 Struts 2 的 Action 按模块配置在多个配置文件中。

3. package 元素

在 Struts 2 框架中使用包来管理 Action，包的作用和 java 中的类包是非常类似的，它主要用于管理一组业务功能相关的 action。在实际应用中，我们应该把一组业务功能相关的 Action 放在同一个包下。

配置包时必须指定 name 属性，如果其他包要继承该包，必须通过该属性进行引用。包的 namespace 属性用于定义该包的命名空间。该属性是可选的，如果不指定该属性，默认的命名空间为“”(空字符串)。

通常每个包都应该继承 struts-default 包，struts－default 包是由 struts 内置的，它定义了 Struts 2 内部的众多拦截器和 Result 类型。

例如在 Struts. xml 配置文件中有如下代码：

```
<struts>
      <package name="commonPackage" extends="struts-default">
...
</struts>
```

Struts 2 很多核心的功能都是通过这些内置的拦截器实现的。如：从请求中把请求参数封装到 action、文件上传和数据验证等等都是通过拦截器实现的。当包继承了 struts-default 包才能使用 Struts 2 提供的这些功能。Struts-default 包是在 struts 2－core－2. x. x. jar 文件中的 struts-default. xml 中定义。struts-default. xml 也是 Struts 2 默认配置文件。Struts 2 每次都会自动加载 struts-default. xml 文件。

包还可以通过 abstract＝“true”定义为抽象包，抽象包中不能包含 action。

与 Java 中的包不同的是，Struts 2 中的包可以扩展另外的包，从而“继承”原有包的所有定义，并可以添加自己包的特有配置，以及修改原有包的部分配置。从这一点上看，Struts 2 中的包更像 Java 中的类。package 有以下几个常用属性：

(1) name：该属性是必选的，指定包的名字，这个名字将作为引用该包的键。

(2) extends：该属性是可选的，允许一个包继承一个或多个先前定义的包。

(3) abstract：该属性是可选的，将其设置为 true，可以把一个包定义为抽象的。

例：

```
<struts>
    <package name="commonPackage" extends="struts-default" abstract="true">
...
</struts>
```

在 Struts 1 中，通过 path 属性指定访问该 action 的 URL 路径。在 Struts 2 中，访问 Struts 2 中的 action 的 URL 路径由两部分组成：包的命名空间 + action 的名称，例如上面例子中访问名为 sum 的 Action 的 URL 路径为：/mystruts/sum（注意：完整路径为：http://localhost：端口/内容路径/mystruts/sum. action）。Action 名称的搜索顺序是：

① 获得请求路径的 URL，例如 url 是：http://server/struts 2/path1/path2/path3/test. action；

② 首先寻找 namespace 为/path1/path2/path3 的 package，如果存在这个 package，则在这个 package 中寻找名字为 test 的 action，如果不存在这个 package 则转步骤③；

③ 寻找 namespace 为/path1/path2 的 package，如果存在这个 package，则在这个 package 中寻找名字为 test 的 action，如果不存在这个 package，则转步骤④；

④ 寻找 namespace 为/path1 的 package，如果存在这个 package，则在这个 package 中寻找名字为 test 的 action，如果仍然不存在这个 package，就去默认的 namespace 的 package 下面去找名字为 test 的 action，如果还是找不到，页面提示找不到 action。

4. Action 元素

表 3-1　Action 元素比较

	Struts 1. x	Struts 2. 0
接口	必须继承 org. apache. struts. action. Action 或者其子类	无须继承任何类型或实现任何接口
表单数据	表单数据封装在 FormBean 中	表单数据包含在 Action 中，通过 Getter 和 Setter 获取

虽然，理论上 Struts 2. 0 的 Action 无须实现任何接口或继承任何类型，但是，为了方便实现 Action，大多数情况下都会继承 com. opensymphony. xwork2. ActionSupport 类，并重载(Override) 此类里的 String execute ( ) 方法。如果没有为 action 指定 class，默认是 ActionSupport。而 ActionSupport 的 execute() 方法默认处理就是返回一个 success 字符串。method 属性用于指定 action 中的那个方法，如果没有指定则默认执行 action 中的 execute() 方法。

当一个请求匹配到某个 Action 名字时，框架就使用这个映射来确定如何处理请求。

```
<action name="struts" class="org. action. StrutsAction">
    <result name="success">/welcome. jsp</result>
    <result name="error">/hello. jsp</result>
</action>
```

如果一个请求要调用 Action 类中除 execute() 方法的其他方法，就需要在 Action 配置中

加以配置。例如,如果在 org. action. StrutsAction 中有另外一个方法为:

```
public class HelloWorldAction{
    private String message;
    ...
    public String execute() throws Exception{
        this.message = "我的第一个 Struts 2 应用";
        return "success";
    }

    public String second() throws Exception{
        this.message = "第二个方法";
        return "success";
    }
}
```

那么如果想要调用这个方法,就必须在 Action 中配置 method 属性,其配置方法为:

```
<!-- name 值是用来和请求匹配的-->
<action name="other " class="org.action.HelloWorldAction " method="second">
    <result name="success">/welcome.jsp</result>
    <result name="error">/hello.jsp</result>
</action>
```

5. result 元素

一个 result 代表一个可能的输出。当 Action 类中的方法执行完成时,返回一个字符串类型的结果代码,框架根据这个结果代码选择对应的 result,向用户输出。

```
<result name ="逻辑视图名" type ="视图结果类型"/>
    <param name ="参数名">参数值</param>
</result>
```

(1) param 中的 name 属性有两个值:

① location:指定逻辑视图。

② parse:是否允许在实际视图名中使用 OGNL 表达式,参数默认为 true。

(2) result 中的 name 属性有如下值:

① success:表示请求处理成功,该值也是默认值。

② error:表示请求处理失败。

③ none:表示请求处理完成后不跳转到任何页面。

④ input:表示输入时如果验证失败应该跳转到什么地方(关于验证后面会介绍)。

⑤ login:表示登录失败后跳转的目标。

(3) type(非默认类型)属性支持的结果类型有以下几种:

① chain:用来处理 Action 链。

② chart:用来整合 JFreeChart 的结果类型。

③ dispatcher：用来转向页面，通常处理 JSP，该类型也为默认类型。

④ freemarker：处理 FreeMarker 模板。

⑤ httpheader：控制特殊 HTTP 行为的结果类型。

⑥ jasper：用于 JasperReports 整合的结果类型。

⑦ jsf：JSF 整合的结果类型。

⑧ redirect：重定向到一个 URL。

⑨ redirect－action：重定向到一个 Action。

⑩ stream：向浏览器发送 InputStream 对象，通常用来处理文件下载，还可用于返回 AJAX 数据。

⑪ tiles：与 Tiles 整合的结果类型。

⑫ velocity：处理 Velocity 模板。

⑬ xslt：处理 XML/XLST 模板。

⑭ plaintext：显示原始文件内容，如文件源代码。

其中，最常用的类型是 dispatcherhe 和 redirect-action 类型。Dispatcherhe 是默认类型，redirect-action 用于当一个 Action 处理结束后，直接将请求重定向到另一个 Action。如下列配置：

```
…
<action name="struts" class="org.action.StrutsAction" >
        <result name="success">/welcome.jsp</result>
        <result name="error">/hello.jsp</result>
</action>
<action name="login" class="org.action.StrutsAction">
        <result name="success" type="redirect－action">struts</result>
</action>
```

§ 3.4　Struts 2 标签库简介

在出现标签库之前，JSP 视图控制和显示技术主要依靠 Java 脚本来实现，JSP 页面重新嵌入了大量的 Java 脚本代码，难以阅读且不同角色协作困难，导致维护成本高。

从 JSP1.1 规范后，JSP 增加了自定义标签库的支持。标签库是一种组件技术，通过标签库，可以将复杂的 Java 脚本代码封装在组件中，开发者只需要使用简单的代码就可以实现复杂的 Java 脚本功能。不仅提供了 Java 脚本的复用性，还提高了开发者的开发效率。Struts 2 标签库相对 struts 1. x 进行了巨大的改进，支持 OGNL 表达式，不再依赖任何表现层技术。

Struts 2 框架本身就提供了丰富的标签库，程序人员也可以自己开发标签库，JSTL 是 Sun 提供的一套标签库，这套标签库的功能非常强大。下面着重介绍框架自身提供的标签库。

3.4.1　Struts 2 标签库概述

1. Struts 2 标签库的组成

Struts 2 框架的标签库可以分为以下三类：

(1) 用户界面标签(UI 标签)：主要用来生成 HTML 元素的标签。

① 表单标签：主要用于生成 HTML 页面的 FORM 元素，以及普通表单元素的标签。

② 非表单标签：主要用于生成页面上的 tree，Tab 页等。

(2) 非用户界面标签(非 UI 标签)：主要用于数据访问，逻辑控制。

① 数据访问标签：主要包含用于输出值栈(ValueStack)中的值，完成国际化等功能的标签。

② 流程控制标签：主要包含用于实现分支，循环等流程控制的标签。

(3) AJAX 标签：用于支持 Ajax 效果

2. Struts 2 标签的使用

struts-tags.tld 文件存在于 struts 2 - core - 2.0.11.jar 压缩文件的 META - INF 目录下，这个文件里定义了 Struts 2 的标签。

要在 jsp 中使用 Struts 2 的标志，先要指明标志的引入。通过 jsp 的代码的顶部加入以下的代码：

```
<%@taglib prefix="s" uri="/struts-tags" %>
```

其中 uri 属性确定标签库的 URI，这个 URI 可以确定一个标签库。而 prefix 属性指定标签库前缀，即所有使用该前缀的标签将由此标签处理。

3.4.2　Struts 2 标签语法

Struts 2 的标签都支持动态数据的访问，标签的属性都可以使用 OGNL 表达式，struts 2 标签的属性具有类型，这些类型可以简单地分为字符串类型和非字符串类型，对于字符串类型的属性，如果要访问动态数据，需要使用%{… }这样的语法，

例如：

```
<s:include value="%{ url }" />
```

Include 标签的 value 属性是字符串类型，Struts 2 将对这个属性进行解析，查找符合%{…}样式的字符串，然后将花括号之间的内容作为 OGNL 表达式进行求值。如果属性中没有%{…}样式的字符串，那么属性的值将被直接看成是字符串数据。

例如：

```
<s:include value="urlTag.action" />
```

对于非字符串类型的属性值，将直接作为 OGNL 表达式进行求值。

例如：

```
<s:property value="username"/>
```

Property标签的value属性是Object类型，它的值username将作为OGNL表达式进行求值，结果是值栈中栈顶的对象的属性名是username的值。如果要为非字符串类型的属性直接指定字符串数据，那么需要使用OGNL中的字符串常量，即用单引号(‘)或双引号(“)将字符串括起来。

例如：

```
<s:property value="'zhangsan'"/>
```

value属性的值‘zhangsan’作为字符串常量，计算结果就是zhangsan，因此输出zhangsan。

除上述用法之外，也可以使用%{…}这样的语法来指定字符串常量。

例如：<s:property value="%{'zhangsan' }"/>

在这种情况下，%{ }将被忽略，花括号中的内容将作为表达式被计算。

总结一下，struts 2标签的属性按照下列的三个规则进行计算。

(1) 所有的字符串属性类型都会解析“%{…}”这样的语法。

(2) 所有的非字符属性类型都不会被解析，而是直接被看作一个OGNL表达式进行求值。

(3) 对于第二个规则的例外情况是，如果非字符串属性使用了“%{…}”语法，那么%{…}将被忽略，花括号中的内容将作为表达式计算。

在使用标签时，如果忘记了某个属性是字符串类型，还是非字符串类型，那么有一个简单的方法，那就是不考虑它是什么类型，统一使用“%{…}”语法。

3.4.3　表单标签

Struts 2的表单标签可以分为两类，form标签本身和单个表单元素的其他标签。Struts 2表单和HTML表单元素基本上是一一对应，Struts 2表单标签包括下列标签：

1. <s:checkboxlist>标签

该标签需要指定一个list属性。用法举例：

```
<form action="">
<s:checkboxlist label="请选择你的爱好"
list="{'sport','sing','dance','read'}" name="hobby">
</s:checkboxlist>
</form>
```

或：

```
<form action="">
<s:checkboxlist  label="请选择你的爱好"
list="#{1:'sport',2:'sing',3:'dance',4:'read'}" name="hobby">
</s:checkboxlist >
</form>
```

这两种方式的区别：前一种根据name取值时取的是选中字符串的值；后一种在页面上显

示的是 value 的值，而根据 name 取值时取的却是对应的 key，这里就是 1、2、3 或 4。

效果如图 3－10 所示：

图 3－10　checkboxlist 标签效果

2. <s:combobox>标签

Combobox 标签生成一个单行文本框和下拉列表框的组合。两个表单元素只能对应一个请求参数，只有单行文本框里的值才包含请求参数，下拉列表框只是用于辅助输入，并没有 name 属性，故不会产生请求参数。用法举例：

```
<form action="">
<s:combobox label="请选择你的爱好"
list="{'sport','sing','dance','read'}" name="hobby">
</s:combobox>
</form>
```

效果如图 3－11 所示。

图 3－11　combobox 标签效果

3. <s:datetimepicker>标签

Datetimepicker 标签用于生成一个日期、时间下拉列表框。当使用该日期、时间列表框选择某个日期、时间时，系统会自动将选中日期、时间输出指定文本框中。注意，使用该标签时需要完成以下几个步骤：

(1) 将 struts2－dojo－plugin－2.1.x.jar 拷贝到/web—inf/lib 下。

(2) 在 jsp 文件中加入<%@ taglib uri="/struts-dojo-tags" prefix="sd"%>和<sd:head/>

具体代码如下：

```
<sd:form action="" method="">
    <s:datetimepicker name="date" label="请选择日期"></s:datetimepicker>
</sd:form>
```

效果如图 3－12 所示。

图 3－12　datetimepicker 标签效果

4. <s:select>标签

Select 标签用于生成一个下拉列表框，通过为该元素指定 list 属性的值，来生成下拉列表框的选项。用法举例：

```
<s:select list="{'sport','dance','sing','swim'}"
label="请选择你的爱好"></s:select>
```

或：

```
<s:select list="hobby" list="#{1:'sport',2:'dance',3:'sing',4:'swim'}"
listKey="key" listValue="value"></s:select>
```

5. <s:radio>标签

Radio 标签的用法与 checkboxlist 用法相似，唯一的区别就是 checkboxlist 生成的是复选框，而 radio 生成的是单选框。用法举例：

```
<s:radio label="性别" list="{'男','女'}" name="sex"></s:radio>
```

或：

```
<s:radio label="性别" list="#{1:'男',0:'女'}" name="sex">
</s:radio>
```

6. <s:head>标签

Head 标签主要用于生成 HTML 页面的 head 部分，在使用<s:datetimepicker>标签时介绍过，要在 head 中加入该标签。因为<s:datetimepicker>标签中有一个日历小控件，其中包含了 JavaScript 代码，所以要在 head 部分加入该标签。

很多表单标签（form 标签除外）的 name 属性和 value 属性之间存在一个独特的关系。name 属性除了为 HTML 表单元素指定名字，在表单提交时作为请求参数的名字外，同时它还映射到 Action 的属性。

一般情况下，name 属性映射到一个简单的 JavaBean 属性，例如 name 属性的值为"operand1"，在表单提交后，struts 2 框架将会调用 Action 的 setOperand1 ()方法来设置属性。

若需在表单元素中显示 Action 属性的数据，就需用到 value 属性。为 value 属性指定表达式"%{operand1}"，这将会调用 Action 的 getOperand1 ()方法，并在表单中显示返回的数据。之后用户可以编辑这个值，然后重新提交它。具体用法可以看本章开始的"数值计算"例子。

3.4.4　非表单标签

非表单标签不经常用到，主要用于在页面中生成一些非表单的可视化元素。下面大致介绍一下这些标签：

(1) a:生成超链接。

(2) actionerror:输出 Action 实例的 getActionError()方法返回的消息。

(3) actionmessage :负责输出 Action 实例的 getActionMessage()方法返回的系列消息。

(4) component:生成一个自定义组件。

(5) div:生成一个 div 片段。

(6) ielderror:输出表单域的类型转换错误、校验错误提示。

(7) tablePanel:生成 HTML 页面的 Tab 页。

(8) tree:生成一个树形结构。

(9) treenode:生成树形结构的节点。

示例：

noFormTagAction. java 文件代码如下：

```
package xz. edu;
import com. opensymphony. xwork2. ActionSupport;
public class noFormTagAction extends ActionSupport
{
    private static final long serialVersionUID = 1L;
@Override
    public String execute()
    {
    //添加两条 Error 信息
        addActionError("第一条错误消息!");
        addActionError("第二条错误消息!");
```

```
        //添加两条普通信息
            addActionMessage("第一条普通消息!");
            addActionMessage("第二条普通消息!");
            return SUCCESS;
        }
}
```

noformtag.jsp 文件代码如下：

```
<%@ page language="java" contentType="text/html; charset=UTF-8"
    pageEncoding="UTF-8"%>
<%@taglib prefix="s" uri="/struts-tags"%>
<%@ taglib prefix="sd" uri="/struts-dojo-tags" %>
<!DOCTYPE html PUBLIC "-//W3C//DTD HTML 4.01 Transitional//EN"
"http://www.w3.org/TR/html4/loose.dtd">
<html>
<head>
<title>form tag</title>
<sd:head/>
</head>
<body>
<s:action name="noFormTag" executeResult="true"/>
<!-- 输出 getActionError()方法返回值 -->
<s:actionerror/>
<!-- 输出 getActionMessage()方法返回值 -->
<s:actionmessage />
</body>
</html>
```

struts.xml 文件代码如下：

```
<?xml version="1.0" encoding="UTF-8"?>
<!DOCTYPE struts PUBLIC
    "-//Apache Software Foundation//DTD Struts Configuration 2.0//EN"
    "http://struts.apache.org/dtds/struts-2.0.dtd">
<struts>
    <package name="struts2" extends="struts-default">
        <action name="noFormTag" class="xz.edu.noFormTagAction">
            <result name="success">/noformtag.jsp</result>
        </action>
    </package>
</struts>
```

noformtag. jsp 文件运行结果如图 3－13 所示。

图 3－13　noformtag. jsp 文件运行结果

3.4.5　数据标签

数据标签属于非 UI 标签，主要用于提供各种数据访问相关的功能，数据标签主要包括以下几个：

(1) property：用于输出某个值。

(2) set：用于设置一个新变量。

(3) param：用于设置参数，通常用于 bean 标签和 action 标签的子标签。

(4) bean：用于创建一个 JavaBean 实例。

(5) action：用于在 JSP 页面直接调用一个 Action。

(6) date：用于格式化输出一个日期。

(7) debug：用于在页面上生成一个调试链接，当单击该链接时，可以看到当前值栈和 Stack Context 中的内容。

(8) il8n：用于指定国际化资源文件的 baseName。

(9) include：用于在 JSP 页面中包含其他的 JSP 或 Servlet 资源。

(10) push：用于将某个值放入值栈的栈顶。

(11) text：用于输出国际化(国际化内容会在后面讲解)。

(12) url：用于生成一个 URL 地址。

1. <s:property>标签

Property 标签的作用是输出指定值。property 标签输出 value 属性指定的值。如果没有指定的 value 属性，则默认输出值栈栈顶的值。该标签有如下几个属性：

(1) default：该属性是可选的，如果需要输出的属性值为 null，则显示 default 属性指定的值。

(2) escape：该属性是可选的，指定是否 escape HTML 代码。

(3) value：该属性是可选的，指定需要输出的属性值，如果没有指定该属性，则默认输出值栈栈顶的值。该属性也是最常用的，如前面用到的：

```
<s:property value="#request.name"/>
```

(4) id:该属性是可选的,指定该元素的标志。

例如,在本章第一个“求和”的例子中的使用该标签的代码如下:

```
<body>
        代数和为非负整数  <br/>
        <h1><s:property value="sum"/></h1>
</body>
```

上述代码中通过 value 获取 Action 实例中的属性 sum 的值。

2. <s:set>标签

Set 标签用于对值栈中的表达式进行求值,并将结果赋给特定作用域中的某个变量名。该标签有如下几个属性:

(1) name:该属性是必选的,重新生成新变量的名字。

(2) scope:该属性是可选的,指定新变量的存放范围。

(3) value:可选,指定将赋给变量的值。如果没指定,则将 ValueStack 栈顶的值赋给新变量。

(4) id:该属性是可选的,指定该元素的引用 id。

下面是一个简单例子,展示了 property 标签访问存储于 session 中的 user 对象的多个字段:

```
<s:property value="#session['user'].username"/>
<s:property value="#session['user'].age"/>
<s:property value="#session['user'].address"/>
```

重复使用#session['user']很麻烦,使用 set 标签定义一个"user"变量,这个变量指向"#session['user'] ",这样使得代码易于阅读:

```
<s:set name="user" value="#session[' user'] " />
<s:property value="#user.username"/>
<s:property value="#user.age" />
<s:property value="#user.address" />
```

3. <s:param>标签

Param 标签主要用于为其他标签提供参数,该标签有如下几个属性:

(1) name:该属性是可选的,指定需要设置参数的参数名。

(2) value:该属性是可选的,指定需要设置参数的参数值。

(3) id:该属性是可选的,指定引用该元素的 id。

例如,要为 name 为 hobby 的参数赋值:

```
<s:param name= "hobby">apple</s:param>
```

或:

```
<s:param name="hobby" value="sport" />
```

上面用法中，指定一个名为 hobby 的参数，该参数的值为 sport 对象的值。如果想指定 hobby 参数的值为 sport 字符串，则应该这样写：

```
<s:param name="hobby" value="'sport'" />
```

4. <s:bean>标签

Bean 标签用于创建一个 JavaBean 的实例。如果要为该 JavaBean 赋值，则需在 JavaBean 类中提供 set 方法。该标签有如下几个属性：

(1) name:该属性是必选的，用来指定要实例化的 JavaBean 的实现类。

(2) id:该属性是可选的，如果指定了该属性，则该 JavaBean 实例会被放入 Stack Context 中，从而允许直接通过 id 属性来访问该 JavaBean 实例。

下面是一个简单的例子。

有一个 Student 类，该类中有 name 属性，并有其 getter 和 setter 方法：

```
public class Student {
      private String name;
      public String getName() {
         return name;
      }
      public void setName(String name) {
         this.name=name;
      }
}
```

然后在 JSP 文件的 body 体中加入下面的代码：

```
<s:bean name="Student">
<s:param name="name" value="'zhangsan'"/>
     <s:property value="name"/>
</s:bean>
```

5. <s:action>标签

使用 action 标签可以允许在 JSP 页面中直接调用 Action。该标签有以下几个属性：

(1) id:该属性是可选的，该属性将会作为该 Action 的引用标志 id。

(2) name:该属性是必选的，指定该标签调用哪个 Action。

(3) namespace:该属性是可选的，指定该标签调用的 Action 所在的 namespace。

(4) executeResult:该属性是可选的，指定是否要将 Action 的处理结果页面包含到本页面。如果值为 true，则包含，如果值为 false，则不包含，默认为 false。

(5) ignoreContextParam:该属性是可选的，指定该页面中的请求参数是否需要传入调用的 Action。如果值为 false，将本页面的请求参数传入被调用的 Action。如为 true，不将本页面的请求参数传入到被调用的 Action。

例：<s:action name="noFormTag" executeResult="true"/>

6. <s:date>标签

Date 标签主要用于格式化输出一个日期。该标签有如下属性：

(1) format:该属性是可选的，如果指定了该属性，将根据该属性指定的格式来格式化日期。

(2) nice:该属性是可选的，该属性的取值只能是 true 或 false，用于指定是否输出指定日期和当前时刻之间的时差。默认为 false，即不输出时差。

(3) name:属性是必选的，指定要格式化的日期值。

(4) id:属性是可选的，指定引用该元素的 id 值。

(5) nice 属性为 true 时，一般不指定 format 属性。因为 nice 为 true 时，会输出当前时刻与指定日期的时差，不会输出指定日期。当没有指定 format，也没有指定 nice="true"时，系统会采用默认格式输出。

用法如下：

```
<s:date name="指定日期取值" format="日期格式"/><!-- 按指定日期格式输出 -->
<s:date name="指定日期取值" nice="true"/><!-- 输出时间差 -->
<s:date name="指定日期取值"/><!--默认格式输出-->
```

7. <s:include>标签

Include 标签用于将一个 JSP 页面或一个 Servlet 包含到本页面中。该标签有如下属性：

(1) value:该属性是必选的，指定需要被包含的 JSP 页面或 Servlet。

(2) id:该属性是可选的，指定该标签的 id 引用。

用法如下：

```
<s:include value="JSP 或 Servlet 文件" id="自定义名称"/>
```

3.4.6　控制标签

用于在呈现结果页面时控制程序的执行流程，根据程序执行的状态输出不同的结果，控制标签包括下列标签：

(1) if:用于控制选择输出的标签。

(2) elseif:用于控制选择输出的标签，必须和 if 标签结合使用。

(3) else:用户控制选择输出的标签，必须和 if 标签结合使用。

(4) append:用于将多个集合拼接成一个新的集合。

(5) generator:用于将一个字符串按指定的分隔符分隔成多个字符串，临时生成的多个子字符串可以使用 iterator 标签来迭代输出。

(6) iterator:用于将集合迭代输出。

(7) merge:用于将多个集合拼接成一个新的集合，但与 append 的拼接方式不同。

(8) sort:用于对集合进行排序。

(9) subset:用于截取集合的部分元素，形成新的子集合。

1. <s:if>/<s:elseif>/<s:else>标签

以上三个标签可以组合使用，但只有 if 标签可以单独使用，而 elseif 和 else 标签必须与 if

标签结合使用。if 标签可以与多个 elseif 标签结合使用，但只能与一个 else 标签使用。其用法格式如下：

```
<s:if test="表达式">
      标签体
</s:if>
<s:elseif test="表达式">
      标签体
</s:elseif>
<!--允许出现多次 elseif 标签-->
      ...
<s:else>
      标签体
</s:else>
```

2. <s:iterator>标签

该标签主要用于对集合进行迭代，这里的集合包含 List、Set，也可以对 Map 类型的对象进行迭代输出。该标签的属性如下：

(1) value：该属性是可选的，指定被迭代的集合，被迭代的集合通常都由 OGNL 表达式指定。如果没有指定该属性，则使用值栈栈顶的集合。

(2) id：该属性是可选的，指定集合元素的 id。

(3) status：该属性是可选的，指定迭代时的 IteratorStatus 实例，通过该实例可判断当前迭代元素的属性。如果指定该属性，其实例包含如下几个方法：

int getCount()：返回当前迭代了几个元素。

int getIndex()：返回当前被迭代元素的索引。

boolean isEven：返回当前被迭代元素的索引元素是否是偶数。

boolean isOdd：返回当前被迭代元素的索引元素是否是奇数。

boolean isFirst：返回当前被迭代元素是否是第一个元素。

boolean isLast：返回当前被迭代元素是否是最后一个元素。

应用举例：

```
<%@ page language="java" pageEncoding="utf-8"%>
<%@taglib uri="/struts-tags" prefix="s" %>
<html>
<head>
   <title>控制标签</title>
</head>
<body>
     <table border="1" width="200">
          <s:iterator value="{'sport','sing','swim','read'}" id="hobby" status="st">
                <tr <s:if test="#st.even">style="background-color:silver"</s:if>>
                      <td><s:property value="hobby"/></td>
```

```
            </tr>
        </s:iterator>
    </table>
</body>
</html>
```

运行结果如图 3－14 所示。

图 3－14　iterator 标签实例运行结果

3. <s:append>标签

该标签用于将多个集合对象拼接起来，组成一个新的集合。

应用举例，可以把上例的 JSP 文件进行修改，其代码为：

```
<%@ page language="java" pageEncoding="utf-8"%>
<%@taglib uri="/struts-tags" prefix="s" %>
<html>
<head>
    <title>控制标签</title>
</head>
<body>
    <s:append id="newList">
    <s:param value="{'sport','sing','swim','read'}"/>
    <s:param value="{'boy','girl'}"/>
    </s:append>
    <table border="1" width="200">
        <s:iterator value="#newList" id="hobby" status="st">
            <tr <s:if test="#st.even">style="background-color:silver"</s:if>>
                <td><s:property value="hobby"/></td>
```

```
            </tr>
        </s:iterator>
    </table>
</body>
</html>
```

运行结果如图 3 - 15 所示。

图 3 - 15　append 标签实例运行结果

4. <s:merge>标签

假设有 2 个集合，第一个集合包含 3 个元素，第二个集合包含 2 个元素，分别用 append 标签和 merge 标签方式进行拼接，它们产生新集合的方式有所区别。

用 append 方式拼接，新集合元素顺序为：

(1) 第 1 个集合中的第 1 个元素；

(2) 第 1 个集合中的第 2 个元素；

(3) 第 1 个集合中的第 3 个元素；

(4) 第 2 个集合中的第 1 个元素；

(5) 第 2 个集合中的第 2 个元素。

用 merge 方式拼接，新集合元素顺序为：

① 第 1 个集合中的第 1 个元素；

② 第 2 个集合中的第 1 个元素；

③ 第 1 个集合中的第 2 个元素；

④ 第 2 个集合中的第 2 个元素；

⑤ 第 1 个集合中的第 3 个元素。

§ 3.5　Struts 2 数据验证

3.5.1　Validate 方法

1. 简介

在 Struts 2 中最简单的验证数据的方法是使用 Validate。通过阅读 ActionSupport 类的源代码，可以看到 ActionSupport 类实现了一个 Validateable 接口，这个接口只有一个 Validate 方法。如果 Action 类实现了这个接口，Struts 2 在调用 execute 方法之前首先会调用这个方法，可以在 Validate 方法中验证，如果发生错误，可以根据错误的 level 选择字段级错误，还是动作级错误。并且可使用 addFieldError 或 addActionError 加入相应的错误信息；如果存在 Action 或 Field 错误，Struts 2 会返回“input”(这个并不用开发人员写，由 Struts 2 自动返回)，如果返回了“input”，Struts 2 就不会再调用 execute 方法了；如果不存在错误信息，Struts 2 在最后会调用 execute 方法。

这两个 add 方法和 ActionErrors 类中的 add 方法类似，只是 add 方法的错误信息需要一个 ActionMessage 对象，比较麻烦。除了加入错误信息外，还可以使用 addActionMessage 方法加入成功提交后的信息。当提交成功后，可以显示这些信息。

以上三个 add 方法都在 ValidationAware 接口中定义，并且在 ActionSupport 类中有一个默认的实现。其实，在 ActionSupport 类中的实现实际上是调用了 ValidationAwareSupport 中的相应的方法，也就是这三个 add 方法是在 ValidationAwareSupport 类中实现的，代码如下：

```
private final ValidationAwareSupport validationAware=new ValidationAwareSupport();
public void addActionError(String anErrorMessage)
{
    validationAware.addActionError(anErrorMessage);
}
public void addActionMessage(String aMessage)
{
    validationAware.addActionMessage(aMessage);
}
public void addFieldError(String fieldName,String errorMessage)
{
    validationAware.addFieldError(fieldName,errorMessage);
}
```

2. 验证实例

下面我们来实现一个简单的验证程序，来体验一个 Validate 方法的使用。先在 Web 根目录建立一个主页面(Validate.jsp)，代码如下：

```
<%@ page language="java" import="java.util.*" pageEncoding="GB18030"%>
<%@taglib prefix="s" uri="/struts-tags"%>
<!DOCTYPE HTML PUBLIC "-//W3C//DTD HTML 4.01 Transitional//EN">
<html>
  <head>
    <title>验证数据</title>
 </head>
 <body>
    <s:actionerror/>
    <s:actionmessage/>
    <s:form action="validate.action" theme="simple">
输入内容:
    <s:textfield name="msg"/>
    <s:fielderror key="msg.hello"/><br/>
    <s:submit/>
    </s:form>
  </body>
</html>
```

在上面的代码中，使用了 Struts 2 的标签：<s:actionerror>、<s:fielderror>和<s:actionmessage>，分别用来显示动作错误信息、字段错误信息和动作信息。如果信息为空，则不显示。

接下来实现一个动作类 validateAction，代码如下：

```
package action;

import com.opensymphony.xwork2.ActionSupport;
public class ValidateAction extends ActionSupport{
        private static final long serialVersionUID = 1L;
        private String msg;
        public String execute()
        {
              System.out.println(SUCCESS);
              return SUCCESS;
        }
        public void validate()
        {
              if(! msg.equalsIgnoreCase("hello"))
              {
                    System.out.println(INPUT);
                    this.addFieldError("msg.hello","必须输入 hello!");
                    this.addActionError("处理动作失败!");
              }
```

```
            else
            {
                  this. addActionMessage("提交成功");
            }
      }
      public String getMsg()
      {
            return msg;
      }
      public void setMsg(String msg)
      {
            this. msg=msg;
      }
}
```

从上面的代码可以看出，Field 错误需要一个 key(一般用来表示是哪一个属性出的错误)，而 Action 错误和 Action 消息只要提供一个信息字符串就可以了。

最后来配置一下 struts. xml 文件，代码如下：

```
<?xml version="1. 0" encoding="UTF-8"?>
<!DOCTYPE struts PUBLIC
   "-//Apache Software Foundation//DTD Struts Configuration 2. 0//EN"
   "http://struts. apache. org/dtds/struts-2. 0. dtd">
   <struts>
   <package name="struts 2" namespace="/mystruts" extends="struts-default">
         <action name="sum1" class="action. FirstAction">
                  <result name="positive">/positive. jsp</result>
                  <result name="negative">/negative. jsp</result>
         </action>
   </package>
   <package name="demo" extends="struts-default">
<action name="validate" class="action. ValidateAction">
<result name="success">/validate. jsp</result>
<result name="input">/validate. jsp</result>
</action>
   </package>
</struts>
```

部署项目到 Tomcat 的 webapps 目录下，启动 Tomcat 之后，打开浏览器，在地址栏输入"http://localhost:8080/ FirstStuts 2/ validate. jsp"，然后回车，浏览器出现如图 3-16 所示界面；如果输入不是"hello"，则出现如图 3-17 所示画面；如果输入"hello"，则出现如图 3-18 所示，即表示运行成功。

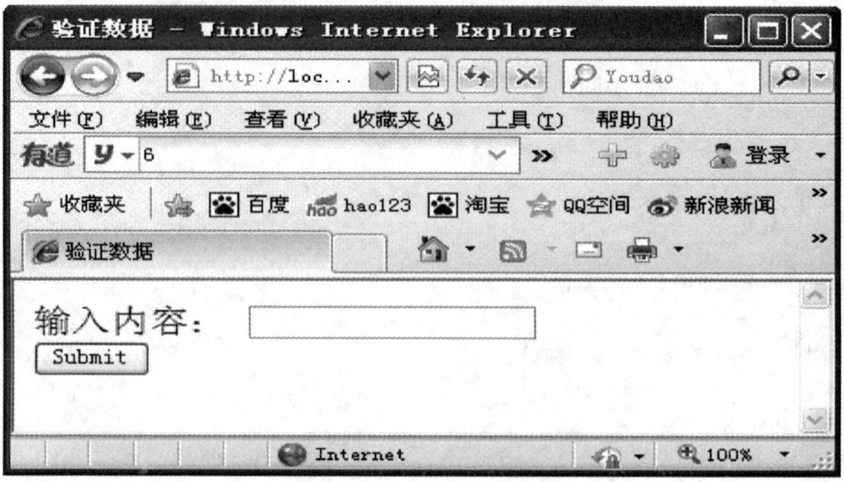

图 3-16　输入验证

图 3-17　验证失败

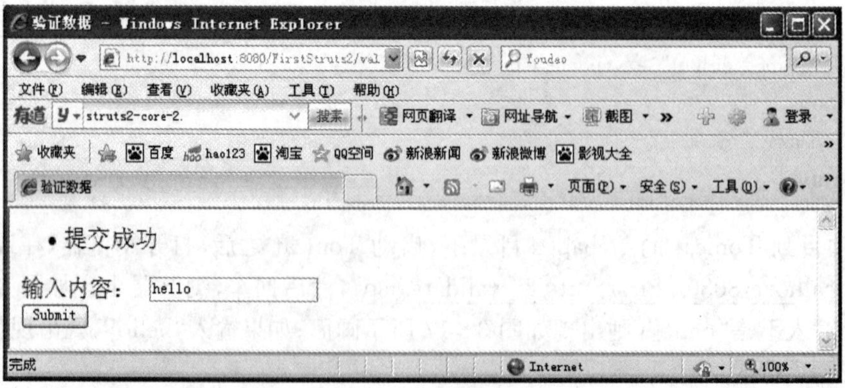

图 3-18　验证成功

3. ValidationAware 接口

还可以使用 ValidationAware 接口的其他方法(由 ValidationAwareSupport 类实现)获得或设置字段错误信息、动作错误信息以及动作消息。如 hasActionErrors 方法判断是否存在动作层的错误,getFieldErrors 获得字段错误信息(一个 Map 对象)。下面是 ValidationAware 接口提供的所有的方法:

```
package com.opensymphony.xwork2;
import java.util.Collection;importjava.util.Map;
public interface ValidationAware
{
        void setActionErrors(Collection errorMessages);
        Collection getActionErrors();
        voids etActionMessages(Collection messages);
        Collection getActionMessages();
        void setFieldErrors(MaperrorMap);
        Map getFieldErrors();
        void addActionError(StringanErrorMessage);
        void addActionMessage(StringaMessage);
        void addFieldError(StringfieldName,StringerrorMessage);
        boolean hasActionErrors();
        boolean hasActionMessages();
        boolean hasErrors();
        boolean hasFieldErrors();
}
```

3.5.2 Validation 框架

上面使用 Validate 方法来验证客户端提交的数据,但使用 Validate 方法会将验证代码和正常的逻辑代码混在一起,这样做并不利于代码维护,而且也很难将这些代码重用于其他程序的验证。Struts 2 提供了一个 Validation 框架,这个框架和 Struts1.x 提供的 Validation 框架类似,也是通过 XML 文件进行配置。

下面将给出一个例子来演示如何使用 Struts 2 的 Validation 框架来进行服务端验证。可按着如下四步来编写这个程序:

1. 建立 Action 类(NewValidateAction.java)

```
package action;


import com.opensymphony.xwork2.ActionSupport;
public class NewValidateAction extends ActionSupport{
        private static final long serialVersionUID = 1L;
        private String msg; //必须输入
        private int age; //在 13 和 20 之间
        public String getMsg()
```

```
    {
        return msg;
    }
    public void setMsg(String msg)
    {
        this.msg=msg;
    }
    public int getAge()
    {
        return age;
    }
    public void setAge(int age)
    {
        this.age=age;
    }
}
```

2. 配置 struts.xml 文件

在 struts 标签中增加一个 package，代码如下：

```
<package name="framedemo" extends="struts-default" >
        <action name="new_validate" class="action.NewValidateAction">
            <result name="input">/validate_form.jsp</result>
            <result name="success">/validate_form.jsp</result>
        </action>
    </package>
```

3. 编写验证规则配置文件

这是一个基于 XML 的配置文件，和 struts1.x 中的 validator 框架的验证规则配置文件类似。但一般放到和要验证的".class"文件在同一目录下，而且配置文件名要使用如下两个规则中的一个来命名：

```
<ActionClassNam>-validation.xml
<ActionClassName>-<ActionAliasName>-validation.xml
```

其中<ActionAliasName>就是 struts.xml 中<ation>的 name 属性值。在本例中使用第一种命名规则，所以文件名是 NewValidateAction-validation.xml。文件的内容如下：

```
<?xml version="1.0" encoding="UTF-8"?>
<!DOCTYPE validators PUBLIC
    "-//OpenSymphonyGroup//XWorkValidator1.0.2//EN"
    "http://www.opensymphony.com/xwork/xwork-validator-1.0.2.dtd">
```

```
<validators>
    <field name="msg">
        <field-validator type="requiredstring">
            <param name="trim">true</param>
            <message>请输入信息</message>
        </field-validator>
    </field>
    <field name="age">
        <field-validator type="int">
            <param name="min">13</param>
            <param name="max">20</param>
            <message>必须在13至20之间</message>
        </field-validator>
    </field>
</validators>
```

这个文件使用了两个规则：requiredstring(必须输入)和int(确定整型范围)。关于其他更详细的验证规则，需访问"http://struts.apache.org/2.0.11.1/docs/validation.html"来查看。

4. 创建JSP页

在Web根目录中建立一个validate_form.jsp文件，代码如下：

```
<%@ page language="java" import="java.util.*" pageEncoding="GB18030"%>
<%@taglib prefix="s" uri="/struts-tags"%>
<!DOCTYPE HTML PUBLIC "-//W3C//DTD HTML 4.01 Transitional//EN">
<html>
  <head>
     <title>数据验证</title>
  </head>
  <body>
   <s:form action="new_validate" >
   <s:textfield name="msg" label="姓名"/>
   <s:textfield name="age" label="年龄"/>
   <s:submit/>
   </s:form>
  </body>
</html>
```

部署项目到Tomcat的webapps目录下，启动Tomcat之后，打开浏览器，在地址栏输入"http://localhost:8080/ FirstStuts 2/validate_form.jsp"，如果输入为空，则出现如图3-19所示画面，即表示验证生效。

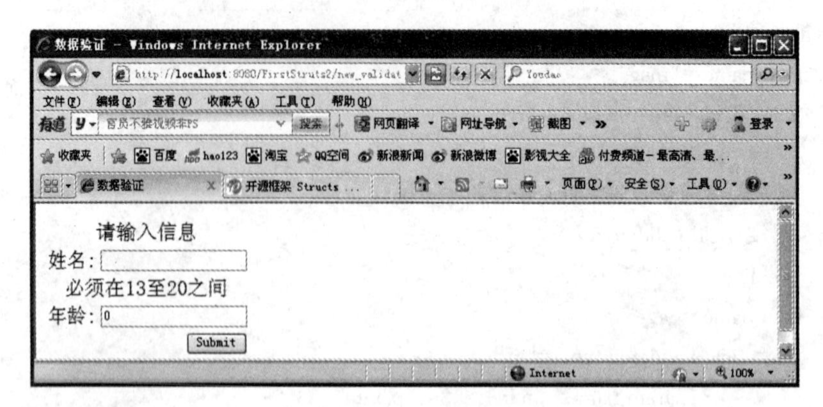

图 3-19　框架验证

§3.6　拦截器

3.6.1　拦截器概述

Struts 2 框架的绝大部分功能是通过拦截器实现的，当 FilterDispatcher 拦截到用户请求后，大量拦截器将会对用户请求进行处理，然后才会调用用户自定义的 Action 类中的方法来处理请求。可见，拦截器是 Struts 2 的核心所在。Struts 2 会首先执行在 struts. xml 文件中引用的拦截器，在执行完所有引用的拦截器的 intercept 方法后，会执行 Action 的 execute 方法。

Struts 2 拦截器类必须从 com. opensymphony. xwork2. interceptor. Interceptor 接口继承，在 Intercepter 接口中有如下三个方法需要实现：

(1) void destroy()；

(2) void init()；

(3) String intercept(ActionInvocation invocation)throws Exception；

其中 intercept 方法是拦截器的核心方法，所有安装的拦截器都会调用这个方法。在 Struts 2 中已经在 struts-default. xml 中预定义了一些自带的拦截器，如 timer、params 等。如果在<package>标签中继承了 struts-default，则当前 package 就会自动拥有 struts-default. xml 中的所有配置。代码如下：

```
<package name="demo" extends="struts-default" > ... </package>
```

在 struts-default. xml 文件中有一个默认的引用，在默认情况下(也就是<action>中未引用拦截器时)会自动引用一些拦截器。这个默认的拦截器引用如下：

```
<default-interceptor-ref name="defaultStack"/>
<interceptor-stack name="defaultStack">
    <interceptor-ref name="exception"/>
```

```
        <interceptor-ref name="alias"/>
        <interceptor-ref name="servletConfig"/>
        <interceptor-ref name="prepare"/>
        <interceptor-ref name="i18n"/>
        <interceptor-ref name="chain"/>
        <interceptor-ref name="debugging"/>
        <interceptor-ref name="profiling"/>
        <interceptor-ref name="scopedModelDriven"/>
        <interceptor-ref name="modelDriven"/>
        <interceptor-ref name="fileUpload"/>
        <interceptor-ref name="checkbox"/>
        <interceptor-ref name="staticParams"/>
        <interceptor-ref name="params">
        <paramname="excludeParams">dojo.. *</param>
        </interceptor-ref>
        <interceptor-ref name="conversionError"/>
        <interceptor-ref name="validation">
         <paramname="excludeMethods">input,back,cancel,browse</param>
        </interceptor-ref>
        <interceptor-ref name="workflow">
        <paramname="excludeMethods">input,back,cancel,browse</param>
        </interceptor-ref>
</interceptor-stack>
```

上面在 defaultStack 中引用的拦截器都可以在<action>中不经过引用就可以使用(如果在<action>中引用了任何拦截器后，要使用在 defaultStack 中定义的拦截器，则需要在<action>中重新引)。

3.6.2 拦截器的使用方法

1. Timer 拦截器

Timer 是 Struts 2 中相对简单的拦截器，这个拦截器对应的类是 com. opensymphony. xwork2. interceptor. TimerInterceptor。它的功能是记录 execute 方法和其他拦截器(在 timer 后面定义的拦截器)的 intercept 方法执行的时间总和。如下面的配置代码所示：

```
<action name="intercept" class="xz. edu. interceptorAction">
  <interceptor-ref name="logger"/>
  <interceptor-ref name="timer"/>
</action>
```

由于在 Timer 后面没有其他的拦截器定义，因此，timer 只能记录 execute 方法的执行时间，在访问 first 动作时，会在控制台输出类似下面的一条信息：

信息：Executed action [/test/first! execute] took 16 ms.

在使用 Timer 拦截器时，需要 commons－logging. jar 的支持。将 logger 引用放到 Timer

的后面，就可以记录 logger 拦截器的 intercept 方法和 Action 的 execute 方法的执行时间总和。代码如下：

```
<action name="intercept" class="xz.edu.interceptorAction">
    <interceptor-ref name="timer"/>
    <interceptor-ref name="logger"/>
</action>
```

大家可以使用如下的 Action 类来测试一下 Timer 拦截器：

```
package xz.edu;
import com.opensymphony.xwork2.ActionSupport;
public class interceptorAction extends ActionSupport
{
    private static final long serialVersionUID = 1L;
@Override
    public String execute() throws Exception{
        Thread.sleep(1000); //延迟 1 秒
        return null;
    }
}
```

Intercept.action 运行输出信息如图 3-20 所示。

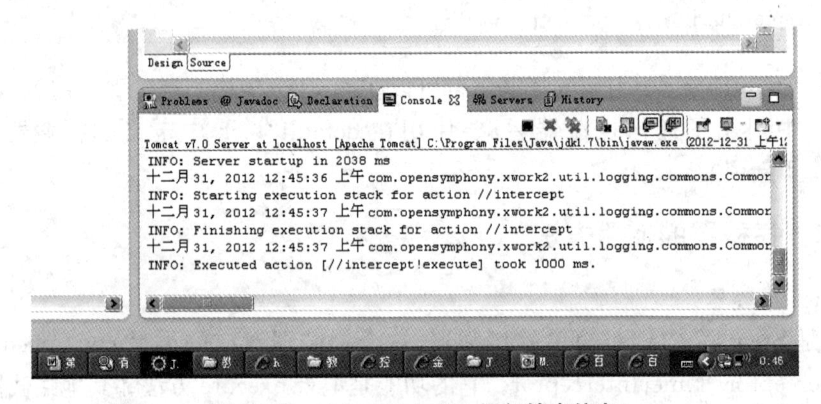

图 3-20　Intercept.action 运行输出信息

2. params 拦截器(通过请求调用 Action 的 setter 方法)

当客户端的一个包含 textfield 的 form 向服务端提交请求时，代码如下：

```
s:form action="first" namespace="/test"
    <s:textfield name="name"/>
    <s:submit/>
</s:form>
```

在提交后，Struts 2 将会自动调用 first 动作类中的 setName 方法，并将 name 文本框中的值通过 setName 方法的参数传入。实际上，这个操作是由 params 拦截器完成的，params 对应的类是 com.opensymphony.xwork2.interceptor.ParametersInterceptor。由于 params 已经在 defaultStack 中定义，因此，在未引用拦截器的<action>中是会自动引用 params 的，如下面的配置代码，在访问 first 动作时，Struts 2 是会自动执行相应的 setter 方法的。

```
<action name="first" class="action.FirstAction">... ...</action>
```

如果在<action>中引用了其他的拦截器，就必须再次引用 params 拦截器，Struts 2 才能调用相应的 setter 方法。如下面的配置代码所示：

```
<action name="first" class="action.FirstAction">
    <interceptor-ref name="timer"/>
    <interceptor-ref name="params"/>
</action>
```

3. static-params 拦截器(通过配置参数调用 Action 的 setter 方法)

static-params 拦截器可以通过配置<params>标签来调用 Action 类的相应的 setter 方法，static-params 拦截器对应的类是 com.opensymphony.xwork2.interceptor.StaticParametersInterceptor。下面配置代码演示了如何使用 static-params 拦截器：

```
<action name="first" class="action.FirstAction">
     <interceptor-ref name="timer"/>
     <param name="who">晓庄学院</param>
     <interceptor-ref name="params"/>
     <interceptor-ref name="static-params"/>
</action>
```

如果 first 动作使用上面的配置，在访问 first 动作时，Struts 2 会自动调用 setWho 方法将“晓庄学院”作为参数值传入 setWho 方法。

4. 使用拦截器栈

为了能在多个动作中方便地引用同一个或几个拦截器，可以使用拦截器栈将这些拦截器作为一个整体来引用。拦截器栈要在 package 标签中使用 interceptors 和子标签 interceptor-stack 来定义。代码如下：

```
package name="demo" extends="struts-default"
interceptors
  interceptor-stack name="mystack"
  interceptor-ref name="timer"/
  interceptor-ref name="logger"/
  interceptor-ref name="params"/
  interceptor-ref name="static-params"/
  /interceptor-stack
/interceptors
action name="first" class="action.FirstAction"
  param name="who"晓庄学院</param>
```

```
  interceptor-ref name="mystack"/
</action>
</package>
```

如上面代码所示，拦截器栈的使用方法与拦截器一样。

3.6.3 实现自定义拦截器

拦截器就是一个普通的 class，只是这个 class 必须实现 com. opensymphony. xwork2. interceptor. Interceptor 接口。Interceptor 接口有如下三个方法：

```
publicinterfaceInterceptorextendsSerializable
{
    Void_destroy();
    Void_init();
    String_intercept(ActionInvocationinvocation)throwsException;
}
```

其中 init 和 destroy 方法只在拦截器加载和释放（都由 Struts 2 自身处理）时执行一次。而 intercept 方法在每次访问动作时都会被调用。Struts 2 在调用拦截器时，每个拦截器类只有一个对象实例，而所有引用这个拦截器的动作都共享这一个拦截器类的对象实例，因此，在实现 Interceptor 接口的类中如果使用类变量，要注意同步问题。

下面将实现一个简单的自定义拦截器，这个拦截器通过请求参数 action 指定一个拦截器类中的方法，并调用这个方法（可以使用这个拦截器对某一特定的动作进行预处理）。如果方法不存在，或是 action 参数不存在，则继续执行下面的代码。

例如，将要访问的 url 如下所示：

http://localhost:8080/book/interceptor. action? action=test

访问上面的 URL 后，拦截器会就会调用拦截器中的 test 方法，如果这个方法不存在，则调用 Invocation. invoke 方法。Invoke 方法和 Servlet 过滤器中调用 FilterChain. doFilter 方法类似，如果在当前拦截器后面还有其他的拦截器，则 Invoke 方法就是调用后面拦截器的 intercept 方法。否则，Invoke 会调用 Action 类的 execute 方法（或其他的执行方法）。

下面我们先来实现一个拦截器的父类 ActionInterceptor。这个类主要实现了根据 action 参数值来调用方法的功能，代码如下：

```
package inteceptor;
import com.opensymphony.xwork2.ActionInvocation;
import com.opensymphony.xwork2.interceptor.Interceptor;
import javax.servlet.http.*;
import org.apache.struts2.*;
public class ActionInterceptor implements Interceptor
{
    private static final long serialVersionUID = 1L;
```

```
    protected final String INVOKE="##invoke";
    public void destroy()
    {
        System.out.println("destroy");
    }
    public void init()
    {
        System.out.println("init");
    }
    public String intercept(ActionInvocation invocation)throws Exception
    {
        HttpServletRequest request=ServletActionContext.getRequest();
        String action=request.getParameter("action");
        System.out.println(this.hashCode());
        if(action!=null)
        {
            try
            {
                java.lang.reflect.Method method=this.getClass().getMethod(action);
                String result=(String)method.invoke(this);
                if(result!=null)
                {
                    if(!result.equals(INVOKE))
                        return result;
                }
                else
                    return null;
            }
            catch(Exception e)
            {
            }
        }
        return invocation.invoke();
    }
}
```

从上面代码中的 intercept 方法可以看出，在调用 action 所指定的方法后，来判断返回值。可能发生的情况有三种：

(1) 返回值为 null，执行 return null；

(2) 返回值为 Invoke，执行 return invockation.invoke()；

(3) 其他情况，执行 return result。result 表示指定方法的返回值，如上面代码所示。

在实现完上面的拦截器父类后，任何继承于 ActionInterceptor 类的拦截器都可以自动根据 action 的参数值调用自身的相应方法。下面我们来实现一个拥有两个动作方法 test 和

print 的拦截器类。代码如下：

```
package interceptor;
import javax. servlet. http. HttpServletResponse;
import org. apache. struts2. ServletActionContext;
public class MultiMethodInterceptor extends ActionInterceptor{
        public String test() throws Exception{
                HttpServletResponse response=ServletActionContext. getResponse();
                response. getWriter(). println("invoke test");
                return this. INVOKE;
                }
        public String print() throws Exception{
                HttpServletResponse response=ServletActionContext. getResponse();
                response. getWriter(). println("invoke print");
                return null;
        }
}
```

test 方法返回了 Invoke，因此，在执行完这个方法后，Struts 2 会接着调用其他拦截器的 intercept 方法或 Action 类的 execute 方法。而 print 方法在执行完后，只是返回了 null，而不再调用其他的方法了，也就是访问如下的 URL 时，动作的 execute 方法将不会执行：

```
http://localhost:8080/book/interceptor. action? action=print
```

下面我们来实现一个 Action 类，代码如下：

```
package xz. edu;
import org. apache. struts2. * ;
import com. opensymphony. xwork2. ActionSupport;
public class myInterceptorAction extends ActionSupport{
        public String other() throws Exception{
                ServletActionContext. getResponse(). getWriter(). println("invoke other");
                return null;
        }
}
```

在这个 Action 类中，只有一个 other 方法，实际上，这个方法相当于 execute 方法，在下面会设置动作的 method 属性为 other。下面我们在 struts. xml 中定义拦截器类和动作，代码如下：

```
<struts>
<package name="demo" extends="struts-default">
        <interceptors>
                <interceptor name="method" class="interceptor. MultiMethodInterceptor"/>
```

```
            <interceptor-stack name="methodStack">
            <interceptor-ref name="method"/>
            <interceptor-ref name="defaultStack"/>
            </interceptor-stack>
        </interceptors>
        <action name="interceptor" class="xz.edu.myInterceptorAction" method="other">
            <interceptor-ref name="methodStack"/>
        </action>
</package>
</struts>
```

在配置上面的 methodStack 拦截器时，要注意在后面引用 defaultStack，否则很多通过拦截器提供的功能将失去。

现在访问如下的 URL：

```
http://localhost:8080/book/interceptor.action? action=test
```

结果如图 3-21 所示。

图 3-21　interceptor.action? action=test 运行结果

而如果访问"http://localhost:8080/book/interceptor.action? action=print"，结果如图 3-22所示。

图 3-22　interceptor.action? action=test 运行结果

可以看出，访问这个 URL 时并没有调用 other 方法。如果随便指定的 action 值的话，则只调用 other 方法。如果访问“http://localhost:8080/book/interceptor.action?action=abc”，就只会输出"invoke other"。

Struts 2 内置的拦截器使用时有很多参数，自定义的拦截器也可以加上同样的参数，如 includeMethods 和 excludeMethods 这两个参数比较常用。其中，includeMethods 指定了拦截器要调用的 Action 类的执行方法(默认是 execute)，即只有在 includeMethods 中指定的方法才会被 Struts 2 调用，而 excludeMethods 恰恰相反，在这个参数中指定的执行方法不会被 Struts 2 调用。如果有多个方法，中间需用逗号","分隔。在 Struts 2 中提供了一个抽象类来处理这两个参数，这个类如下：

```
com.opensymphony.xwork2.interceptor.MethodFilterInterceptor
```

继承于这个类的拦截器类都会自动处理 includeMethods 和 excludeMethods 参数，如下面的拦截器类所示：

```
package interceptor;
import com.opensymphony.xwork2.ActionInvocation;
import com.opensymphony.xwork2.interceptor.*;
public class MyFilterInterceptor extends MethodFilterInterceptor{
        private String name;
        public String getName(){
                return name;
        }
        public void setName(String name){
                this.name=name;
        }
        @override protected String doIntercept(ActionInvocation invocation) throws Exception
        {
                System.out.println("doIntercept");
                System.out.println(name);
                return invocation.invoke();
        }
}
```

MethodFilterInterceptor 的子类需要实现 doIntercept 方法(相当于 Interceptor 的 intercept 方法)，如上面代码所示。在上面的代码中还有一个 name 属性，是为了读取拦截器的 name 属性而设置的，如下面的配置代码所示：

```
<?xml version="1.0"encoding="UTF-8"?>
<!DOCTYPEstrutsPUBLIC
        "-//ApacheSoftwareFoundation//DTDStrutsConfiguration2.0//EN"
        "http://struts.apache.org/dtds/struts-2.0.dtd">
        <struts>
                <package name="demo" extends="struts-default" namespace="/test">
```

```
            <interceptors>
                <interceptor name="method" class="interceptor.
MultiMethodInterceptor"/>
                <interceptor name="filter"  class="interceptor.
MyFilterInterceptor">
                    <param name="includeMethods">other</param>
                    <param name="name">china</param>
            </interceptor>
            <interceptor-stack name="methodStack">
                <interceptor-ref name="method"/>
                <interceptor-ref name="filter"/>
                <interceptor-ref name="defaultStack"/>
                </interceptor-stack>
            </interceptors>
            <action name="interceptor"  class="xz.edu.myInterceptorAction"  method="
other">
                <interceptor-ref name="methodStack"/>
            </action>
        </package>
    </struts>
```

再次访问"http://localhost:8080/ book/interceptor action? action=test"，Struts 2 就会调用 MyFilterInterceptor 的 doIntercept 方法来输出 name 属性值。如果将上面的 includeMethods 参数值中的 other 去掉，则 Action 类的 other 方法不会被执行。

§3.7　文件上传

3.7.1　上传单个文件

上传文件是很多 Web 程序都具有的功能。在 Struts 1. x 中已经提供了用于上传文件的组件，而在 Struts 2 中提供了一个更为容易操作的上传文件组件。所不同的是，Struts1. x 的上传组件需要一个 ActionForm 来传递文件，而 Struts 2 的上传组件是一个拦截器(这个拦截器不用配置，是自动装载的)。接下来先介绍用 Struts 2 上传单个文件的操作方法，最后介绍用 Struts 2 上传任意多个文件的操作方法。

用 Struts 2 实现上传单个文件的功能非常容易实现，只要使用普通的 Action 即可。但为了获得一些上传文件的信息，如上传文件名、上传文件类型以及上传文件的 Stream 对象，就需要按着一定规则来为 Action 类增加一些 getter 和 setter 方法。

在 Struts 2 中，用于获得和设置 java. io. File 对象(Struts 2 将文件上传到临时路径，并使用 java. io. File 打开这个临时文件)的方法是 getUpload 和 setUpload。获得和设置文件名的方法是 getUploadFileName 和 setUploadFileName，获得和设置上传文件内容类型的方法是 getUploadContentType 和 setUploadContentType。Struts 2 的文件上传默认使用的是

Common-FileUpload 文件上传组件。因此，需要在 Web 应用中增加两个 Jar 包，即 common-io-1. 3. 2. jar 和 commons-fileupload-1. 1. 1. jar，下面是用于上传的动作类的完整代码。

uploadAction. java 文件代码如下：

```
package xz.edu;
import java.io.*;
import com.opensymphony.xwork2.ActionSupport;
public class UploadAction extends ActionSupport
{
    private File upload;
    private String fileName;
    private String uploadContentType;
    public String getUploadFileName()
    {
        return fileName;
    }
    public void setUploadFileName(String fileName)
    {
        this.fileName=fileName;
    }
    public File getUpload()
    {
        return upload;
    }
    public void setUpload(File upload)
    {
        this.upload=upload;
    }
    public void setUploadContentType(String contentType)
    {
        this.uploadContentType=contentType;
    }
    public String getUploadContentType()
    {
        return this.uploadContentType;
    }
    public String execute() throws Exception
    {
        java.io.InputStream is=new java.io.FileInputStream(upload);
        java.io.OutputStream os=new java.io.FileOutputStream("d:\\upload\\"+fileName);
//该路径根据实际情况设置
        byte buffer[]=new byte[8192];
        int count=0;
```

```
            while((count=is.read(buffer))>0)
            {
                os.write(buffer,0,count);
            }
            os.close();
            is.close();
            return SUCCESS;
        }
}
```

在 execute 方法中的实现代码很简单，把临时文件复制到指定的路径（在这里是 d:\upload）中。上传文件的临时目录的默认值是 javax.servlet.context.tempdir 的值，但可以通过 struts.properties（和 struts.xml 在同一个目录下）的 struts.multipart.saveDir 属性设置。Struts 2 上传文件的默认大小限制是 2M，也可以通过 struts.properties 文件中的 struts.multipart.maxSize 修改，如 struts.multipart.maxSize=2048，则表示一次上传文件的总大小不能超过 2k 字节。

上传文件 upload.jsp 的代码如下：

```
<%@page language="java" import="java.util.*" pageEncoding="GBK"%>
<%@taglib prefix="s" uri="/struts-tags"%>
<html>
    <head>
        <title>上传单个文件</title>
    </head>
    <body>
        <s:form action="upload" namespace="/test" enctype="multipart/form-data">
            <s:file name="upload" label="输入要上传的文件名"/>
            <s:submit value="上传"/>
        </s:form>
    </body>
</html>
```

也可以在 upload.jsp 页中通过<s:property>获得文件的属性（文件名和文件内容类型），代码如下：

```
<s:property value="uploadFileName"/>
```

upload.jsp 文件运行结果如图 3－23 所示。

图 3－23　upload.jsp 文件运行结果

3.7.2　上传多个文件

在 Struts 2 中，上传任意多个文件比较容易实现。首先，要想上传任意多个文件，需要在客户端使用 DOM 技术生成任意多个标签。name 属性值都相同。

uploadMore.jsp 文件代码如下：

```
<html>
    <head>
        <script language="javascript">
            function addComponent()
            {
                var uploadHTML = document.createElement("<input type='file'name=
'upload'/>");
                document.getElementById("files").appendChild
(uploadHTML);
                uploadHTML=document.createElement("<p/>");
                document.getElementById("files").appendChild
(uploadHTML);
            }
        </script>
    </head>
    <body>
        <input type="button" onclick="addComponent();"value="添加文件"/>
        <br/>
```

```
        <form onsubmit="return true;"action="/uploadMore.action"
            method="post" enctype="multipart/form-data">
            <span id="files">
                <input type='file'name='upload'/>
                <p/>
            </span>
            <input type="submit"value="上传"/>
        </form>
    </body>
</html>
```

上面的 javascript 代码可以生成任意多个<input type="file ">标签，name 的值都为 file(须注意的是，上面的 javascript 代码只适用于 IE 浏览器，firefox 等其他浏览器需要使用他的代码)。至于 Action 类，和上传单个文件的 Action 类基本一致，只需要将三个属性的类型改为 List 即可。

UploadMoreAction.java 文件代码如下：

```
package xz.edu;
import java.io.*;
import com.opensymphony.xwork2.ActionSupport;
public class UploadMoreAction extends ActionSupport
{
    private java.util.ListFile uploads;
    private java.util.ListString fileNames;
    private java.util.ListString uploadContentTypes;
    public java.util.ListString getUploadFileName()
    {
        return fileNames;
    }
    public void setUploadFileName(java.util.List<String> fileNames)
    {
        this.fileNames=fileNames;
    }
    public java.util.List<File> getUpload()
    {
        return uploads;
    }
    public void setUpload(java.util.List<File> uploads)
    {
        this.uploads=uploads;
    }
    public void setUploadContentType(java.util.List<String> contentTypes)
    {
```

```
        this.uploadContentTypes=contentTypes;
    }
    public java.util.List<String> getUploadContentType()
    {
        return this.uploadContentTypes;
    }
    public String execute() throws Exception
    {
        if(uploads! =null)
        {
            int i=0;
            for(;i<uploads.size();i++)
            {
                java.io.InputStream is=new java.io.FileInputStream(uploads.get(i));
                java.io.OutputStream os= new java.io.FileOutputStream("d:upload"+
fileNames.get(i));
                byte buffer[]=new byte[8192];
                int  count=0;
                while((count=is.read(buffer))>0)
                {
                    os.write(buffer,0,count);
                }
                os.close();
                is.close();
            }
        }
        return SUCCESS;
    }
}
```

在 execute 方法中，只是对 List 对象进行枚举，在循环中的代码和上传单个文件时的代码基本相同。如使用过 struts1.x 的上传组件的人，会感觉 Struts 2 的上传功能更容易实现，在 Struts1.x 中上传多个文件时，需要建立带索引的属性的。而在 Struts 2 中，则很简单。文件 uploadMore.jsp 文件运行界面如图 3-24 所示。

图 3-24　uploadMore.jsp 文件运行结果

巩固练习

1. 深入理解并写出 Struts 2 程序开发的基本配置。
2. 为 Action 类增加第二个方法并调用。
3. append、iterator 等标签的使用。
4. 在书中上传例子基础上完成下载功能。

第 4 章　Hibernate 和 MyBatis

学习目标

1. 了解 Hibernate 运行原理。
2. 掌握 Hibernate 实际运用。
3. 了解 MyBatis 运行原理。
4. 掌握 MyBatis 实际运用。
5. 了解 Hibernate 与 MyBatis 联系和区别。

§4.1　ORM 简介

对象/关系映射 ORM(Object-Relation Mapping)是用于将对象与对象之间的关系对应到数据库表与表之间的关系的一种模式。简单地说,ORM 是通过使用描述对象和数据库之间映射的元数据,将 Java 程序中的对象自动持久化到关系数据库中。对象和关系数据是业务实现的两种表现形式,业务实体在内存中表现为对象,在数据库中表现为关系数据。

内存中的对象之间存在着关联和继承关系。而在数据库中,关系数据无法直接表达多对多关联和继承关系。因此,ORM 系统一般以中间件的形式存在,主要实现程序对象到关系数据库数据的映射。一般的 ORM 包括四个部分:对持久类对象进行 CRUD 操作的 API、用来规定类和类属性相关查询的语言或 API、规定 mapping metadata 的工具以及可以让 ORM 实现同事务对象一起进行 dirty checking、lazy association fetching 和其他优化操作的技术。

§4.2　Hibernate 体系结构

Hibernate 作为模型层/数据访问层。它通过配置文件(hibernate. cfg. xml 或 hibernate. properties)和映射文件(*. hbm. xml)把 Java 对象或持久化对象(Persistent Object,PO)映射到数据库中的数据表,然后通过操作 PO,对数据库中的表进行各种操作,其中 PO 就是 POJO(普通 Java 对象)加映射文件。Hibernate 的体系结构如图 4-1 所示。

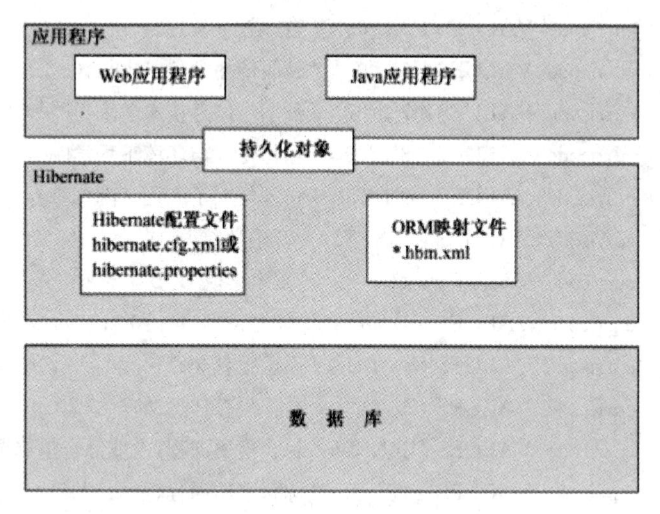

图4-1 Hibernate的体系结构

§4.3 Hibernate应用实例

开发Hibernate项目的步骤如下：

1. 建立数据库及表

本书使用Mysql数据库首先创建名为"jwxt"数据库，并在jwxt数据库中建立学科代码(curricode)表，其表结构创建及插入记录代码如下。

```
-- ----------------------------
-- Table structure for 'curricode'
-- ----------------------------
DROP TABLE IF EXISTS 'curricode';
CREATE TABLE 'curricode' (
   'ID' int(11) NOT NULL AUTO_INCREMENT,
   'Code' varchar(2) NOT NULL,
   'Name' varchar(50) NOT NULL,
   'Type' int(11) NOT NULL,
   'ParentID' int(11) NOT NULL,
   PRIMARY KEY ('ID')
) ENGINE=InnoDB AUTO_INCREMENT=81 DEFAULT CHARSET=utf8;
INSERT INTO 'curricode' VALUES ('1', '01', '人文学院', '0', '0');
INSERT INTO 'curricode' VALUES ('2', '02', '教育科学学院', '0', '0');
INSERT INTO 'curricode' VALUES ('3', '03', '外国语学院', '0', '0');
```

```
INSERT INTO 'curricode' VALUES ('4', '04', '经济与管理学院', '0', '0');
INSERT INTO 'curricode' VALUES ('5', '05', '教师教育学院', '0', '0');
INSERT INTO 'curricode' VALUES ('6', '06', '新闻传播学院', '0', '0');
INSERT INTO 'curricode' VALUES ('7', '07', '数学与信息技术学院', '0', '0');
INSERT INTO 'curricode' VALUES ('8', '08', '生物化工与环境工程学院', '0', '0');
INSERT INTO 'curricode' VALUES ('9', '09', '物理与电子工程学院', '0', '0');
INSERT INTO 'curricode' VALUES ('10', '10', '音乐学院', '0', '0');
INSERT INTO 'curricode' VALUES ('11', '11', '美术学院', '0', '0');
INSERT INTO 'curricode' VALUES ('12', '12', '体育系', '0', '0');
INSERT INTO 'curricode' VALUES ('13', '1', '通识教育必修课程', '2', '0');
INSERT INTO 'curricode' VALUES ('14', '2', '通识教育任选课程', '2', '0');
INSERT INTO 'curricode' VALUES ('16', '5', '综合素质课程(专业综合选修课)', '2', '0');
INSERT INTO 'curricode' VALUES ('17', '6', '综合素质课程(素质拓展)', '2', '0');
INSERT INTO 'curricode' VALUES ('18', '3', '学科基础课程', '2', '0');
INSERT INTO 'curricode' VALUES ('19', '4', '专业课程', '2', '0');
```

2. 在 MyEclipse 中创建对 SQL Server 的连接

启动 MyEclipse，选择【Window】→【Open Perspective】→【MyEclipse Database Explorer】菜单项，打开 MyEclipse Database 浏览器，右击菜单，如图 4－2 所示。

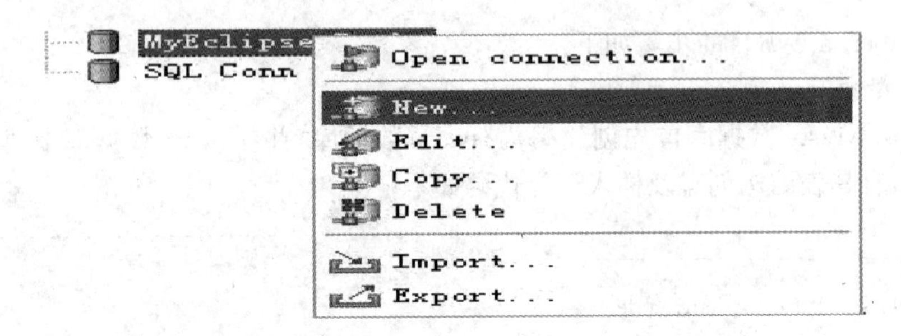

图 4－2　MyEclipse 设置数据库连接

选择【New…】菜单项，出现如图 4－3 所示的对话框，编辑数据库连接驱动。

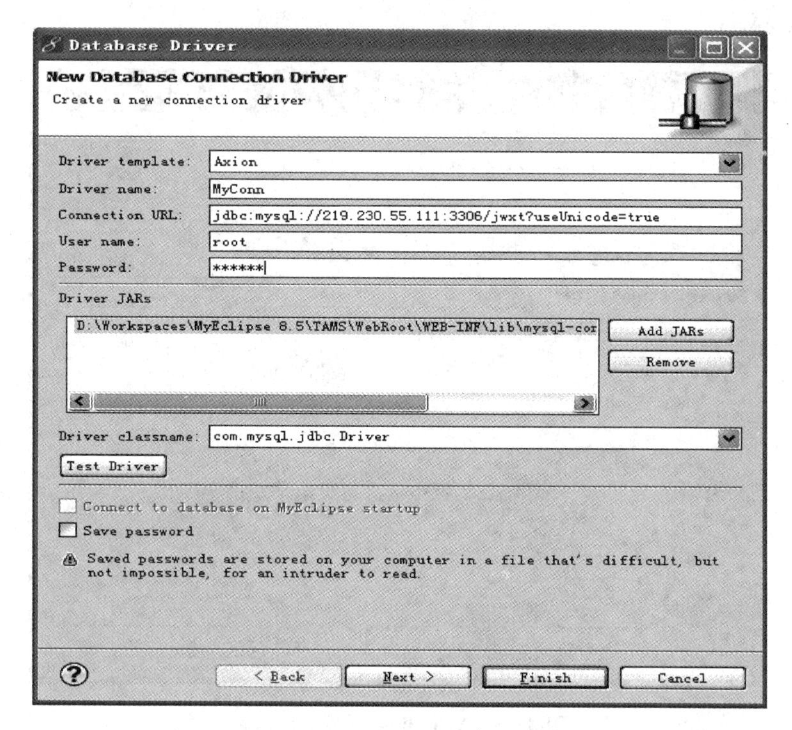

图 4-3 Myeclipse 设置连接参数

3. 创建 Web 项目，命名为"HibernateTest"

4. 添加 Hibernate 开发能力

右击项目名"HibernateTest"，选择【MyEclipse】→【Add Hibernate Capabilites】菜单项，出现如图 4-4 所示的对话框，选择 Hibernate 框架应用版本及所需要的类库。

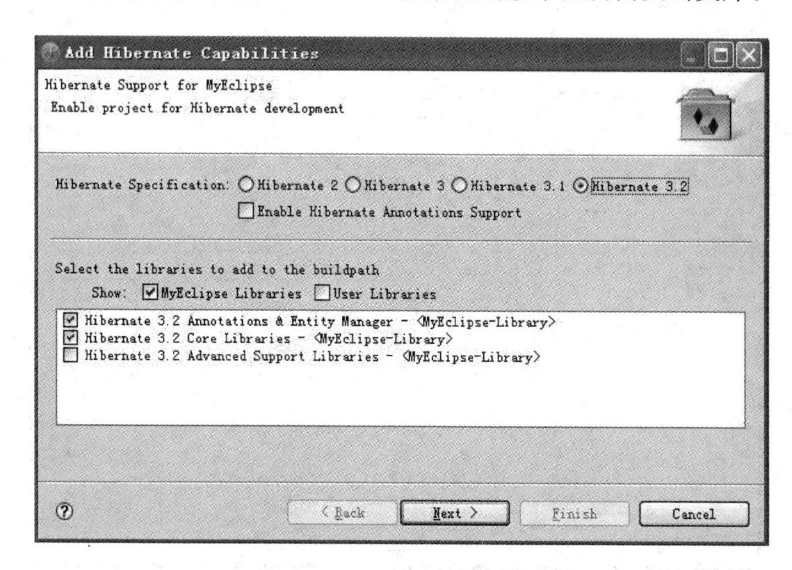

图 4-4 MyEclipse 添加类库

单击【Next】按钮，进入如图 4-5 所示界面。创建 Hibernate 配置文件 hibernate.cfg.xml 文件，将该文件放在 src 文件夹下，后面会详细介绍该文件内容。这里先说明添加 Hibernate

开发功能的步骤。

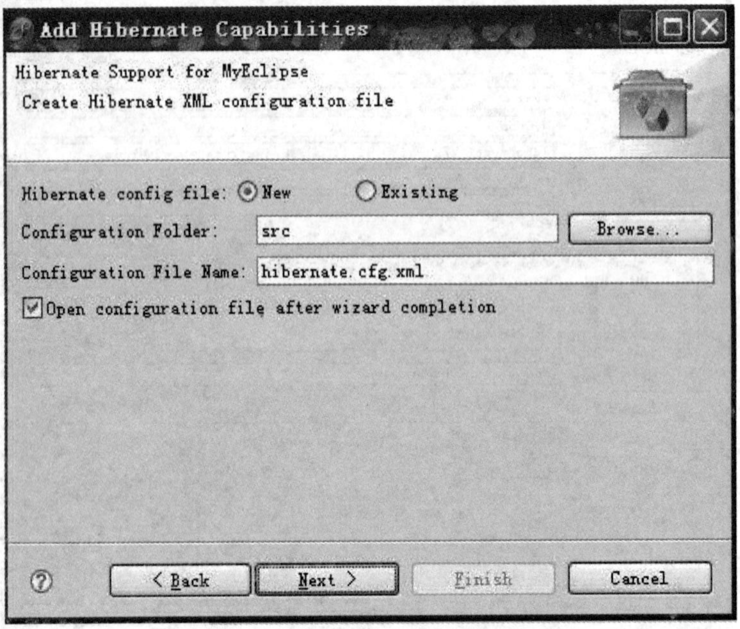

图 4－5　MyEclipse 配置 Hibernate

单击【Next】按钮，进入如图 4－6 所示界面，指定 Hibernate 数据库连接细节。由于在前面已经配置一个名为 MyConn 的数据库连接，所以这里只需要选择 DB Driver 为“MyConn”即可。

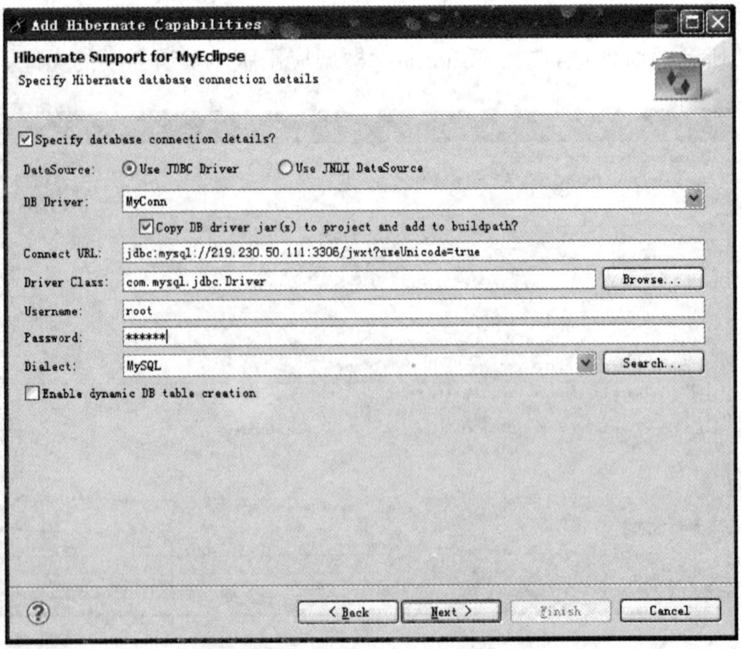

图 4－6　配置 Hibernate 连接参数

单击【Next】按钮，出现如图 4－7 所示界面。

Add Hibernate Capabilities
Create Hibernate SessionFactory for MyEclipse
Define SessionFactory properties
Create SessionFactory class?
Java source folder: src　New...
Java package: org.util　Browse...　New...
Class name: HibernateSessionFactory
Java Compliance Level: Java 1.4　Java 5
< Back　Next >　Finish　Cancel

图 4－7　添加 Hibernate 类

Hibernate 中有一个与数据库打交道重要的类 Session。而这个类是由工厂 SessionFactory 创建的。这个界面询问是否需要创建 SessionFactory 类。如果需要创建，还需要指定创建的位置和类名。这些接口都会在后面详细介绍。单击【Finish】按钮，完成 Hibernate 的配置。

5. 生成数据库表对应的 Java 类对象和映射文件

首先在 MyEclispse 下创建一个名为“org. model”的包，这个包将用来存放与数据库表对应的 Java 类 POJO。

从主菜单栏，选择【Windows】→【Open Perspective】→【Other】→【MyEclipse Database Explorer】菜单项，打开 MyEclipse Database Explorer 视图。打开前面创建的 MyConn 数据连接，选择【XSCJ】→【dbo】→【TABLE】菜单项，右击 KCB 表，选择【Hibernate Reverse Engineering…】菜单项，如图 4－8 所示，将启动 Hibernate Reverse Engineering 向导，该向导用于完成从已有的数据库表生成对应的 Java 类和相关映像文件的配置工作。

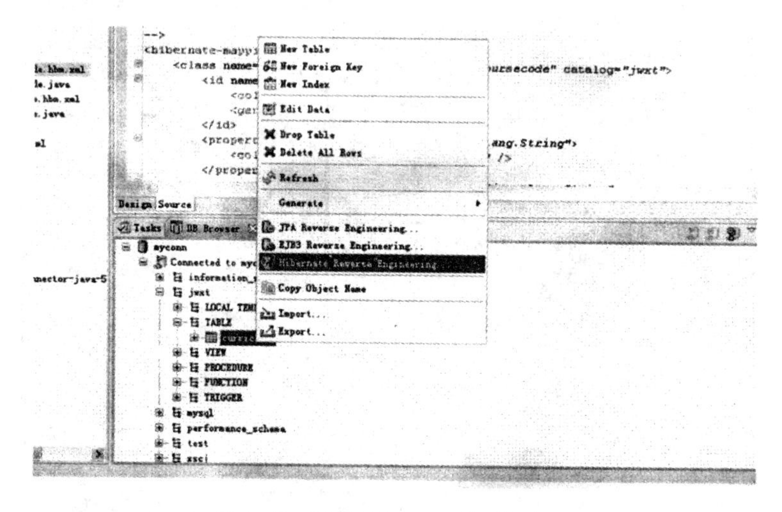

图 4－8　利用 Hibernate 生成 JOPO

选择生成的 Java 类和映像文件所在的位置，如图 4－9 所示。POJO(Plain Old Java

Object,简单的 Java 对象),通常也称为 VO(Value Object,值对象)。

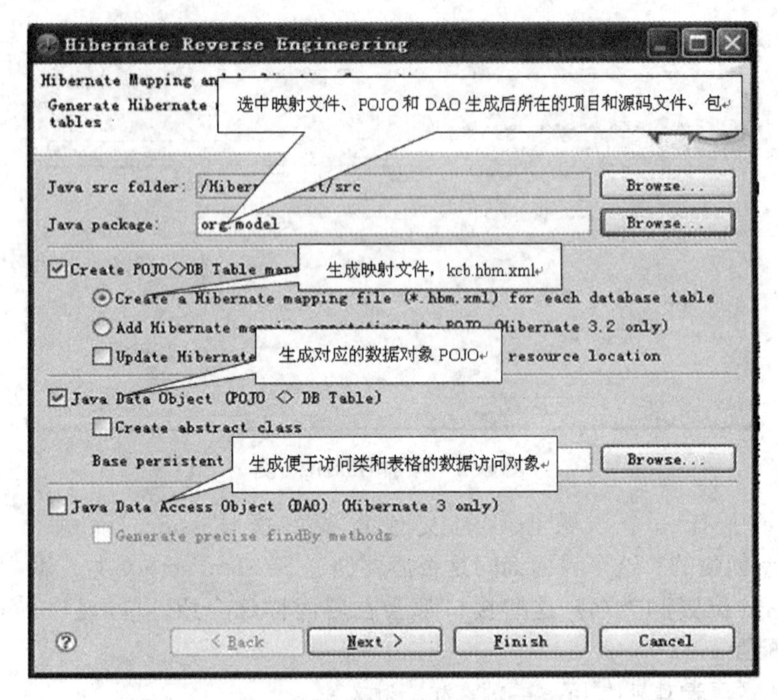

图 4-9 设置生成 JOPO 参数

使用 POJO 名称是为了避免和 EJB 混淆起来,其中有一些属性及 getter、setter 方法。当然,如果有一个简单的运算属性也是可以的,但不允许有业务方法。单击【Next】按钮,进入如图 4-10 所示的界面,选择主键生成策略。

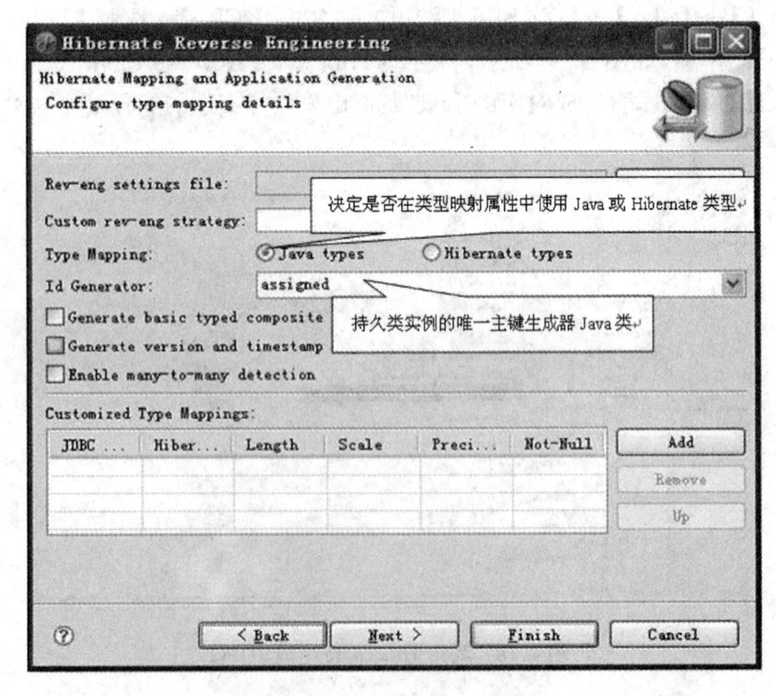

图 4-10 设置生成 JOPO 策略

6. 创建测试类

在 src 文件夹下创建包 test，在该包下建立测试类，命名为 Test. java，其代码如下：

```
package test;
import java. util. List;
import org. hibernate. Query;
import org. hibernate. Session;
import org. hibernate. Transaction;
import org. model. Curricode;
import org. util. HibernateSessionFactory;
public class Test {
    public static void main(String[] args) {
     // 调用 HibernateSessionFactory 的 getSession 方法创建 Session 对象
     Session session=HibernateSessionFactory. getSession();
     // 创建事务对象
     Transaction ts=session. beginTransaction();
     Curricode Cc=new Curricode ();                              // 创建 POJO 类对象
     Cc. setCode ("22");                              // 设置课程号
     Cc.  setType (0);                           // 设置课程名
     Cc.  setName (“物电学院”);                  // 设置开学学期
     // 保存对象
     session. save(Cc);
     ts. commit();                                    // 提交事务
     Query query=session. createQuery("from Curricode where code=22");
     List list=query. list();
     Curricode Cc1=(Curricode) list. get(0);
     System. out. println(Cc1. getName());
     HibernateSessionFactory. closeSession();          // 关闭 Session
    }
}
```

7. 运行

该程序为 Java Application，可以直接运行。运行程序后，控制台就会打印出“物电学院”。在完全没有操作数据库的情况下，就完成了对数据的插入，结果如图 4 - 11 所示。

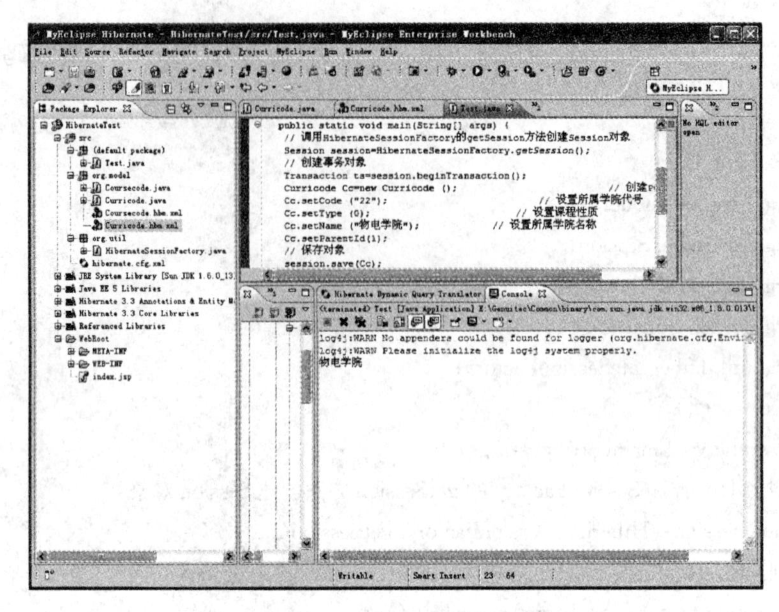

图 4-11　Hibernate 运行结果

§4.4　Hibernate 文件作用详解

1. POJO 类

POJO 类如下：

```
package org. model;

/* *
 * Curricode entity. @author MyEclipse Persistence Tools
 */

public class Curricode implements java. io. Serializable {
    // Fields
    private Integer id;
    private String code;
    private String name;
    private Integer type;
    private Integer parentId;
    // Constructors
    /* * default constructor */
    public Curricode() {
    }
    /* * full constructor */
```

```
    public Curricode(Integer id, String code, String name, Integer type,
            Integer parentId) {
        this.id = id;
        this.code = code;
        this.name = name;
        this.type = type;
        this.parentId = parentId;
    }
    // Property accessors
    public Integer getId() {
        return this.id;
    }
    public void setId(Integer id) {
        this.id = id;
    }
    public String getCode() {
        return this.code;
    }
    public void setCode(String code) {
        this.code = code;
    }
    public String getName() {
        return this.name;
    }
    public void setName(String name) {
        this.name = name;
    }
    public Integer getType() {
        return this.type;
    }
    public void setType(Integer type) {
        this.type = type;
    }
    public Integer getParentId() {
        return this.parentId;
    }
    public void setParentId(Integer parentId) {
        this.parentId = parentId;
    }
}
```

可以发现，该类中的属性和表中的字段是一一对应的。那么通过什么方法把它们一一映射起来呢？这里就是 Curridcode. hbm. xml 文件配置表与 POJO 类的映射关系，其代码如下：

```
<?xml version="1.0" encoding="utf-8"?>
<!DOCTYPE hibernate-mapping PUBLIC "-//Hibernate/Hibernate Mapping DTD 3.0//EN"
"http://hibernate.sourceforge.net/hibernate-mapping-3.0.dtd">
<!--
        Mapping file autogenerated by MyEclipse Persistence Tools
-->
<hibernate-mapping>
    <class name="org.model.Curricode" table="curricode" catalog="jwxt">
        <id name="id" type="java.lang.Integer">
            <column name="ID" />
            <generator class= "identity" />
        </id>
        <property name="code" type="java.lang.String">
            <column name="Code" length="2" not-null="true" />
        </property>
        <property name="name" type="java.lang.String">
            <column name="Name" length="50" not-null="true" />
        </property>
        <property name="type" type="java.lang.Integer">
            <column name="Type" not-null="true" />
        </property>
        <property name="parentId" type="java.lang.Integer">
            <column name="ParentID" not-null="true" />
        </property>
    </class>
</hibernate-mapping>
```

2. 文件配置

(1) 类、表映射配置

```
<class name="org.model.Curricode" table="curricode"catalog="jwxt">
```

name 属性指定 POJO 类为 org.model.Curricode，table 属性指定当前类对应数据库表 curricode。

(2) id 映射配置

```
<id name="id" type="java.lang.Integer">
            <column name="ID" />
            <generator class="identity" />
</id>
```

Hibernate 的主键生成策略分为三大类：Hibernate 对主键 id 赋值、应用程序自身对 id 赋值、由数据库对 id 赋值。

identity：采用数据库提供的主键生成机制，当设置<generator class="identity"/>时，则应用 SQL Server、MySQL 中的自增主键生成机制。

（3）属性、字段映射配置

属性、字段映射将映射类属性与库表字段相关联。

```
<property name="code" type="java.lang.String">
        <column name="Code" length="2" not-null="true" />
    </property>
```

name="code" 指定映像类中的属性名为“code”，此属性将被映像到指定的库表字段 Code。type="java.lang.String"指定映像字段的数据类型。column name="Code"指定类的 code 属性映射 Curricode 表中的 Code 字段。

（4）Hibernate.cfg.xml 文件

该文件是 Hibernate 重要的配置文件，该文件是由 MyEclipse 中的 Hibernate 设置后自动生成，其主要包含代码及解释如下所示。

```
<?xml version='1.0' encoding='UTF-8'?>
<!DOCTYPE hibernate-configuration PUBLIC
        "-//Hibernate/Hibernate Configuration DTD 3.0//EN"
        "http://hibernate.sourceforge.net/hibernate-configuration-3.0.dtd">

<!-- Generated by MyEclipse Hibernate Tools.                    -->
<hibernate-configuration>

    <session-factory>
        <property name="dialect">
            org.hibernate.dialect.MySQLDialect
        </property>
        <!-- 数据库连接地址 -->
        <property name="connection.url">
            jdbc:mysql://127.0.0.1:3306/jwxt? useUnicode=true
        </property>
        <!-- 数据库连接用户名 -->
        <property name="connection.username">root</property>
        <!-- 数据库连接密码 -->
        <property name="connection.password">123456</property>
        <!-- 数据库连接驱动 -->
        <property name="connection.driver_class">
            com.mysql.jdbc.Driver
        </property>
        <property name="myeclipse.connection.profile">MyConn</property>
        <!--表与 POJO 类对应的映射文件   -->
        <mapping resource="org/model/Teachercode.hbm.xml" />
```

```
        <mapping resource="org/model/Curricode.hbm.xml" />
    </session-factory>
</hibernate-configuration>
```

（5）HibernateSessionFactory

HibernateSessionFactory 类是自定义的 SessionFactory，可自定义命名。这里用的是 HibernateSessionFactory，略去其具体代码和注释。

在 Hibernate 中，Session 负责完成对象持久化操作。该文件负责创建 Session 对象，以及关闭 Session 对象。从该文件可以看出，Session 对象的创建大致需要以下三个步骤：

初始化 Hibernate 配置管理类 Configuration；通过 Configuration 类实例创建 Session 的工厂类 SessionFactory；通过 SessionFactory 得到 Session 实例。

§4.5 Hibernate 核心接口

4.5.1 Configuration 接口

Configuration 负责管理 Hibernate 的配置信息。Hibernate 运行时需要一些底层实现的基本信息。这些信息包括：数据库 URL、数据库用户名、数据库用户密码、数据库 JDBC 驱动类、数据库 dialect。用于对特定数据库提供支持，其中包含了针对特定数据库特性的实现，如 Hibernate 数据库类型到特定数据库数据类型的映射等。

使用 Hibernate 首先必须提供这些基础信息以完成初始化工作，为后续操作做好准备。这些属性在 Hibernate 配置文件 hibernate.cfg.xml 中加以设定，当调用 Configuration config =new Configuration().configure()时，Hibernate 会自动在目录下搜索 hibernate.cfg.xml 文件，并将其读取到内存中作为后续操作的基础配置。

4.5.2 SessionFactory 接口

SessionFactory 负责创建 Session 实例，可以通过 Configuration 实例构建 SessionFactory。

```
Configuration config=new Configuration().configure();
SessionFactory sessionFactory=config.buildSessionFactory();
```

Configuration 实例 config 会根据当前的数据库配置信息，构造 SessionFacory 实例并返回。SessionFactory 一旦构造完毕，即被赋予特定的配置信息。也就是说，之后 config 的任何变更将不会影响到已经创建的 SessionFactory 实例 sessionFactory。如果需要使用基于变更后的 config 实例的 SessionFactory，需要从 config 重新构建一个 SessionFactory 实例。

SessionFactory 保存了对应当前数据库配置的所有映射关系，同时也负责维护当前的二级数据缓存和 Statement Pool。由此可见，SessionFactory 的创建过程非常复杂、代价高昂。这也意味着，在系统设计中充分考虑到 SessionFactory 的重用策略。由于 SessionFactory 采用了线程安全的设计，可由多个线程并发调用。

4.5.3　Session 接口

Session 是 Hibernate 持久化操作的基础，提供了众多持久化方法，如 save、update、delete 等。通过这些方法，透明地完成对象的增加、删除、修改、查找等操作。

同时，须注意的是，Hibernate Session 的设计是非线程安全的，即一个 Session 实例同时只可由一个线程使用。同一个 Session 实例的多线程并发调用将导致难以预知的错误。

Session 实例由 SessionFactory 构建：

```
Configuration config=new Configuration(). configure();
SessionFactory sessionFactory=config. buldSessionFactory();
Session session=sessionFactory. openSession();
```

4.5.4　Transaction 接口

Transaction 是 Hibernate 中进行事务操作的接口，Transaction 接口是对实际事务实现的一个抽象，这些实现包括 JDBC 的事务、JTA 中的 UserTransaction，甚至可以是 CORBA 事务。之所以这样设计是可以让开发者能够使用一个统一的操作界面，使得自己的项目可以在不同的环境和容器之间方便地移值。事务对象通过 Session 创建。例如以下语句：

```
Transaction ts=session. beginTransaction();
```

4.5.5　Query 接口

在 Hibernate 2. x 中，find()方法用于执行 HQL 语句。Hibernate 3. x 废除了 find()方法，取而代之的是 Query 接口，它们都用于执行 HQL 语句。Query 和 HQL 是分不开的。

```
Query query=session. createQuery("from Kcb where kch=198");
```

例如以下语句：

```
Query query=session. createQuery("from Kcb where kch=?");
```

就要在后面设置其值：

```
Query. setString(0, "要设置的值");
```

上面的方法是通过"?"来设置参数，还可以用":"后跟变量的方法来设置参数，如上例可以改为：

```
Query query=session. createQuery("from Kcb where kch=:kchValue");
Query. setString("kchValue","要设置的课程号值");
```

其使用方法是相同的，例如：

```
Query. setParameter(0, "要设置的值");
```

Query 还有一个 list()方法，用于取得一个 List 集合的示例，此示例中包括可能是一个 Object 集合，也可能是 Object 数组集合。例如：

```
Query query=session.createQuery("from Kcb where kch=198");
List list=query.list();
```

§ 4.6 HQL

HQL 是 Hibernate Query Language 的缩写。HQL 是面向对象的查询语句，它的语法和 SQL 语句有些相似，在运行时才得以解析。HQL 并不像 SQL 那样是数据表操作语言，HQL 是面向对象的操作语言，HQL 总的对象名是区分大小写的，HQL 中操作的是对象而不是元素，并且支持多态。其主要通过 Query 来操作一创建方式。

Hibernate 现在的最高版本是 3.x，Hibernate 3 不支持 insert 语句，只支持 update、delete 和 select 三种 hql 语句。保存的时候可以使用 Hibernate 封装对象，也可以使用 session 对象。insert 可以通过 session 对象，通过表与 POJO 类的映射关系插入记录到数据表中，具体操作可参考本章第一个例子。

1. HQL 查询

下面介绍 HQL 的几种常用的查询方式。

查询是 HQL 中最常用的一种数据操作方式。下面以学院编码信息为例说明几种查询情况。

(1) 查询所有学院信息

```
...
Session session=HibernateSessionFactory.getSession();
Transaction ts=session.beginTransaction();
Query query=session.createQuery("from Curricode");
List list=query.list();
ts.commit();
HibernateSessionFactory.closeSession();
```

(2) 查询学院编码是'22'的信息

```
...
Session session=HibernateSessionFactory.getSession();
Transaction ts=session.beginTransaction();
Query query=session.createQuery("select name from Curricode where code=22");
List list=query.list();
ts.commit();
HibernateSessionFactory.closeSession();
```

(3) 使用范围运算查询

```
…
Session session=HibernateSessionFactory.getSession();
Transaction ts=session.beginTransaction();
// 查询这样的学院信息,编码在 0~4 之间
Query query=session.createQuery("from Curricode where (code between 0 and 4);
List list=query.list();
ts.commit();
HibernateSessionFactory.closeSession();
```

2. HQL 更新与删除

以下利用 HQL 进行实体更新和删除的技术。这项技术功能是 Hibernate 3 的新加入的功能,在 Hibernate 2 中是不具备的。

(1) 把编码'22'更新为'23'

```
…
Session session=HibernateSessionFactory.getSession();
Transaction trans=session.beginTransaction();
String hql="update Curricode curr set curr.code=23 where curr.code=22";
Query queryupdate=session.createQuery(hql);
queryupdate.executeUpdate();
trans.commit();
```

通过这种方式可以在 Hibernate 3 中一次性完成批量数据的更新。

(2) 删除编码是'23'的所有记录

```
…
Session session=HibernateSessionFactory.getSession();
Transaction trans=session.beginTransaction();
String hql="delete from Curricode curr where curr.code=23";
Query queryupdate=session.createQuery(hql);
queryupdate.executeUpdate();
trans.commit();
```

§4.7 Hibernate 关系映射

4.7.1 环境准备

Hibernate 关系映射的主要任务是实现数据库关系表与持久化类之间的映射,本节主要讲述 Hibernate 关系映射的几种关联关系。

1. 一对一(Person-IdCard)

(1) 基于主键的 one-to-one(person 的映射文件)

关联关系 ER 图如图 4-12 所示。

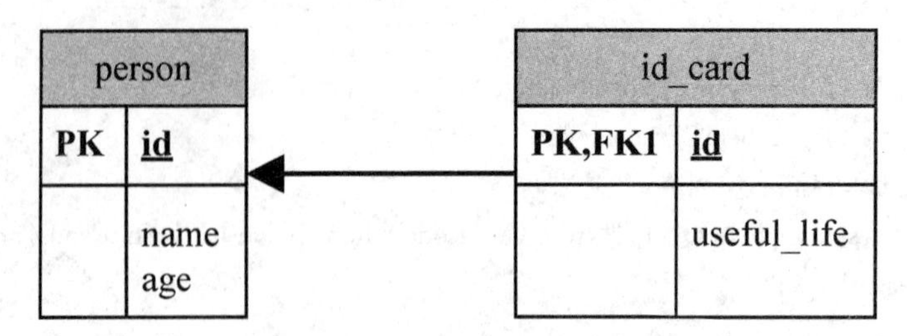

图 4-12　一对一(基于主键)

```
<id name="id">
    <generator class="foreign"><param name="property">idCard</param></generator>
<id>
<one-to-one name="idCard" constrained="true"/>
```

(2) 基于外健的 one-to-one,可以描述为多对一,加 unique=“true”约束

关联关系 ER 图如图 4-13 所示。

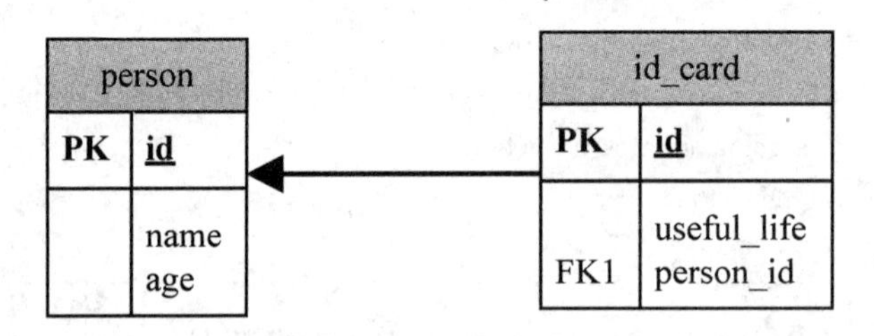

图 4-13　一对一(基于外键)

```
<one-to-one name="idCard" property-ref="person"/>
    property-ref 用于指定关联类的一个属性,这个属性将会和本外键相对应
<many-to-one name="person" column="person_id" unique="true" not-null="true"/>
                <!-唯一的多对一,其实就便成了一对一了-->
```

2. 多对一(Employee-Department)

关联关系 ER 图如图 4-14 所示。

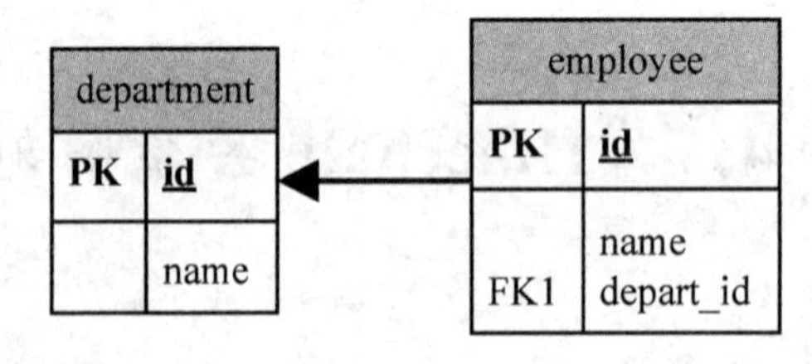

图 4-14　多对一

```
映射文件<many-to-one name="depart" column="depart_id"/>
```

3. 一对多(Department-Employee)

```
<set name="employees">
        <key column="depart_id"/>
        <one-to-many class="Employee"/>
</set>
```

4. 多对多(teacher-student)

关联关系 ER 图如图 4-15 所示。

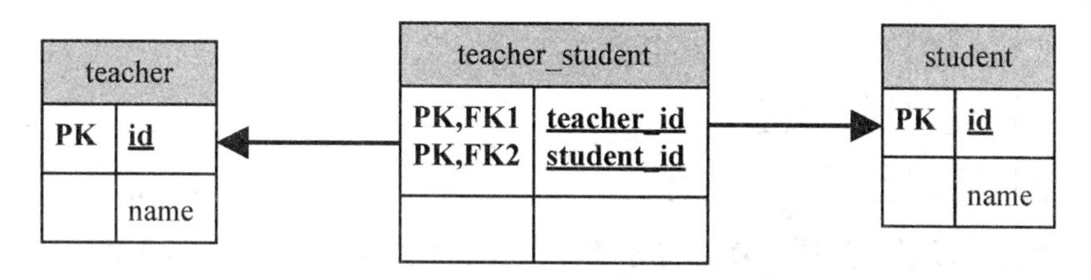

图 4-15　多对多

多对多在操作和性能方面都不太理想,所以多对多的映射使用较少,实际使用中最好转换成一对多的对象模型。Hibernate 会创建中间关联表,转换成两个一对多。

```
<set name="teacher" table="teacher_student">
        <key column="teacher_id"/>
        <many-to-many class="Student" column="student_id"/>
</set>
```

下面就以多对多为例,介绍在 Hibernate 中操作关联关系。

4.7.2　具体实例

学生和教师就是多对多的关系,一个学生可以选择多门课程,因此可以有多个教师,而一个教师又可以教多个学生。多对多关系在关系数据库中不能直接实现,还必须依赖一张连接表。如表 4-1、表 4-2 和表 4-3 所示。

表 4-1　学生表(student)

学段名称	数据类型	主　键	自　增	允许为空	描　述
ID	int	是	增 1		ID 号
SNUMBER	varc har(10)				学号
SNAME	varc har(10)			是	姓名
SAGE	int			是	年龄

表 4-2　教师表(teacher)

学段名称	数据类型	主　键	自　增	允许为空	描　述
ID	int	是	增 1		ID 号
TNUMBER	varc har(10)				教师工号
TNAME	varc har(10)			是	姓名

表 4-3　连接表(tea-stu)

学段名称	数据类型	主　键	自　增	允许为空	描　述
SID	int	是			学生 ID 号
TID	varc har(10)	是			教师 ID 号

这里以学生选择了多门课程就选择了多个教师为例实现多对多单向关联。步骤如下：

① 在项目 Hibernate_mapping 的 org. model 包下编写生成数据库表对应的 Java 类对象和映射文件。

student 表对应的 POJO 类如下：

```
package org. model;
import java. util. HashSet;
import java. util. Set;
public class Student implements java. io. Serializable{
    private int id;
    private String snumber;
    private String sname;
    private int sage;
    private Set teachers=new HashSet();
    //省略上述各属性的 getter 和 setter 方法
}
student 表与 Student 类的 ORM 映射文件 Student. hbm. xml 代码如下：
<?xml version="1. 0" encoding="utf-8"?>
<!DOCTYPE hibernate-mapping PUBLIC "-//Hibernate/Hibernate Mapping DTD 3. 0//EN"
"http://hibernate. sourceforge. net/hibernate-mapping-3. 0. dtd">
<hibernate-mapping>
    <class name="org. model. Student" table="student">
            <id name="id" type="java. lang. Integer">
                <column name="ID" length="4" />
                <generator class="identity"/>
        </id>
        <property name="snumber" type="java. lang. String">
            <column name="SNUMBER"/>
```

```
        </property>
        <property name="sname" type="java.lang.String">
                <column name="SNAME" />
        </property>
        <property name="sage" type="java.lang.Integer">
                <column name="SAGE"></column>
        </property>
        <set name="teachers"    // set 标签表示此属性为 Set 集合类型，由 name 指定属性名称
          table="tea_stu"       // 连接表的名称
          lazy="true"       // 表示此关联为延迟加载，所谓延迟加载就是到了用的时候进行
                            // 加载，避免大量暂时无用的关系对象
          cascade="all">// 级联程度
          <key column="SID"></key>   // 指定参照 student 表的外键名称
          <many-to-many class="org.model.Teacher"      // 被关联的类的名称
                    column="TID"/>           // 指定参照 course 表的外键名称
        </set>
    </class>
</hibernate-mapping>
```

Teacher 表对应的 POJO 如下：

```
package org.model;
public class Teacher implements java.io.Serializable{
    private int id;
    private String tnumber;
    private String tname;
    //省略上述各属性的 getter 和 setter 方法。
}
teacher 表与 Teacher 类的 ORM 映射文件 Teacher.hbm.xml。
<?xml version="1.0" encoding="utf-8"?>
<!DOCTYPE hibernate-mapping PUBLIC "-//Hibernate/Hibernate Mapping DTD 3.0//EN"
"http://hibernate.sourceforge.net/hibernate-mapping-3.0.dtd">
<hibernate-mapping>
    <class name="org.model.Teacher" table="teacher">
        <id name="id" type="java.lang.Integer">
            <column name="ID" length="4" />
            <generator class="identity"/>
        </id>
        <property name="tnumber" type="java.lang.String">
            <column name="TNUMBER"/>
        </property>
        <property name="tname" type="java.lang.String ">
            <column name="TNAME" />
```

```
            </property>
        </class>
</hibernate-mapping>
```

② 在 Hibernate. cfg. xml 文件中加入如下的配置映射文件的语句。

```
<mapping resource="org/model/Student. hbm. xml"/>
<mapping resource="org/model/Teacher. hbm. xml"/>
```

③ 编写测试代码。

在 src 文件夹下的包 test 的 Test 类中加入代码如下所示：

```
...
Teacher teacher1=new Teacher();
Teacher teacher2=new Teacher();
Teacher teacher3=new Teacher();
teacher1. setTnumber("101");
teacher1. setTname("王芳");
teacher2. setTnumber("102");
teacher2. setTname("李明");
teacher3. setTnumber("103");
teacher3. setTname("张宏");
Set teachers=new HashSet();
teachers. add(teacher 1);
teachers. add(teacher 2);
teachers. add(teacher 3);
Student stu=new Student();
stu. setSnumber("081101");
stu. setSname("李方方");
stu. setSage(21);
stu. setTeachers(teachers);
session. save(stu);
//设置完成后可通过 Session 对象调用 session. save(stu)完成持久化
...
```

§4.8 MyBatis 简介及应用

MyBatis 的前身是 iBatis，iBatis 本是由 Clinton Begin 开发，后来捐给 Apache 基金会，成立了 iBatis 开源项目。2010 年 5 月该项目由 Apahce 基金会迁移到了 Google Code，并且改名为 MyBatis。接下来首先简要了解 iBatis 的相关知识。

4.8.1 iBatis 简介

相对 Hibernate 和 Apache OJB 等“一站式”ORM 解决方案而言，iBatis 是一种“半自动化”的 ORM 实现。这里的“半自动化”是相对于 Hibernate 等提供了全面的数据库封装机制的“全自动”ORM 实现而言，“全自动”ORM 实现了 POJO 和数据库表之间的映射，以及 SQL 的自动生成和执行。而 iBatis 的着力点，则在于 POJO 与 SQL 之间的映射关系。也就是说，iBatis 并不会为程序员在运行期自动生成 SQL 执行。具体的 SQL 需要程序员编写，然后通过映射配置文件，将 SQL 所需的参数以及返回的结果字段映射到指定 POJO。

使用 iBatis 提供的 ORM 机制，对业务逻辑实现人员而言，面对的是纯粹的 Java 对象，这一层与通过 Hibernate 实现 ORM 而言基本一致，对于具体的数据操作，Hibernate 会自动生成 SQL 语句。而 iBatis 则要求开发者编写具体的 SQL 语句，相对 Hibernate 等“全自动”ORM 机制而言，iBatis 以 SQL 开发的工作量和数据库移植性上的让步，为系统设计提供了更大的自由空间。作为“全自动”ORM 实现的一种有益补充，iBatis 的出现显得别具意义。

4.8.2 MyBatis 简介

MyBatis 是支持普通 SQL 查询，存储过程和高级映射的优秀持久层框架。MyBatis 消除了几乎所有的 JDBC 代码和参数的手工设置以及对结果集的检索。MyBatis 可以使用简单的 XML 或注解用于配置和原始映射，将接口和 Java 的 POJO(Plain Old Java Objects，普通的 Java 对象)映射成数据库中的记录。

MyBatis 是一个数据持久层(ORM)框架，把实体类和 SQL 语句之间建立了映射关系，是一种半自动化的 ORM 实现。MyBatis 的优点：

(1) 基于 SQL 语法，简单易学；

(2) 能了解底层组装过程；

(3) SQL 语句封装在配置文件中，便于统一管理与维护，降低了程序的耦合度；

(4) 程序调试方便。

1. 与传统 JDBC 的比较

① 减少了 61%的代码量；

② 最简单的持久化框架；

③ 架构级性能增强；

④ SQL 代码从程序代码中彻底分离，可重用；

⑤ 增强了项目中的分工；

⑥ 增强了移植性。

2. JDBC 与 MyBatis 直观对比

Jave 类使用 MyBatis 连接操作数据库代码如下：

```
Class.forName("com.mysql.jdbc.Driver");
Connection conn=DriverManager.getConnection(url,user,password);
java.sql.PrepareStatement st=conn.prepareStatement(sql);
st.setInt(0,1);
st.execute();
```

```
java.sql.ResultSet rs=st.getResultSet();
while(rs.next()){
string result=rs.getString(colname);
}
<mapper namespace="org.mybatis.example.BlogMapper">
 <select id="selectBlog" parameterType="int" resultType="Blog">
     select * form Blog where id=#{id}
 </select>
</mapper>
```

MyBatis 就是将上面这几行代码分解包装，机制如图 4-16 所示。

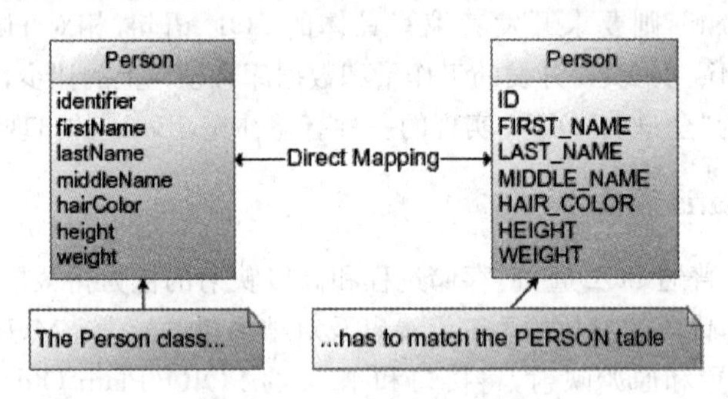

图 4-16 MyBatis 工作机制

接下来，对上面代码进行分析如下：

① 前两行是对数据库的数据源的管理包括事务管理；

② 3、4 两行 MyBatis 通过配置文件来管理 SQL 以及输入参数的映射；

③ 6、7、8 行 MyBatis 获取返回结果到 Java 对象的映射，也是通过配置文件管理。

3. 与 Hibernate 的对比

(1) Hibernate 的映射关系：

MyBatis

- 它是一个 SQL 语句映射的框架(工具)；
- 注重 POJO 与 SQL 之间的映射关系，不会为程序员在运行期自动生成 SQL；
- 自动化程度低、手工映射 SQL，灵活程度高；
- 需要开发人员熟炼掌据 SQL 语句。

Hibernate

- 主流的 ORM 框架、提供了从 POJO 到数据库表的全套映射机制；
- 会自动生成全套 SQL 语句；
- 因为自动化程度高、映射配置复杂，api 也相对复杂，灵活性低；
- 开发人员不必关注 SQL 底层语句开发。

(2) MyBatis 的映射关系，如图 4-17 所示。

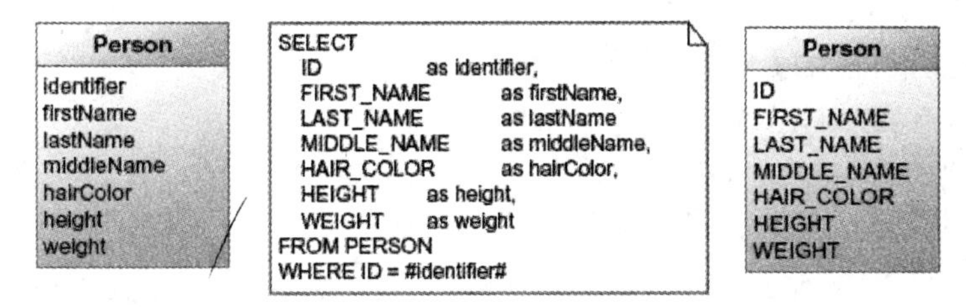

图 4 - 17　MyBatis 映射关系

4.8.3　MyBatis 工作流程

工作流程如图 4 - 18 所示。

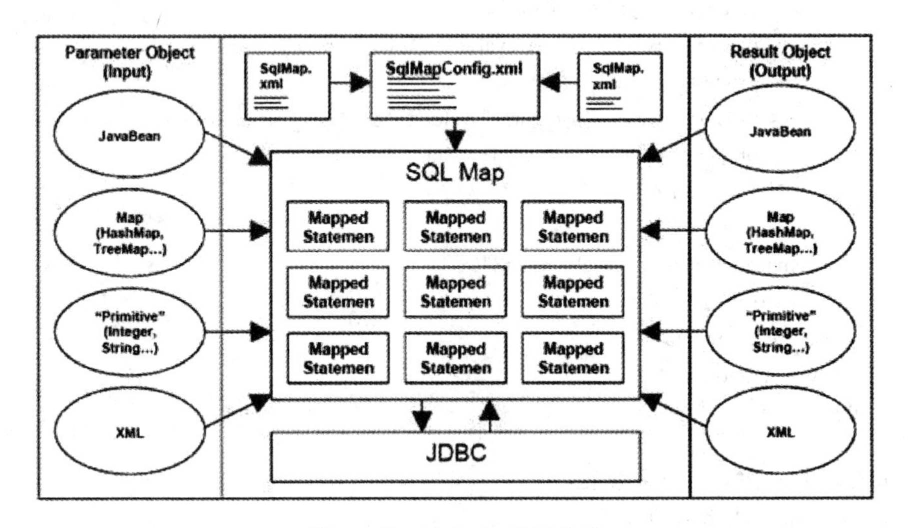

图 4 - 18　MyBatis 映射关系

4.8.4　MyBatis 基本要素

(1) configuration. xml 全局配置文件

(2) mapper. xml 核心映射文件

(3) SqlSession 接口

1. configuration. xml 文件简介

该文件是系统的核心配置文件，包含数据源和事务管理器等设置和属性信息，XML 文档结构如下：

```
configuration 配置；
    properties 可以配置在 Java 属性配置文件中；
    settings 修改 MyBatis 在运行时的行为方式；
    typeAliases  为 Java 类型命名一个短的名字；
    typeHandlers 类型处理器；
    objectFactory 对象工厂；
```

```
        plugins 插件；
        environments 环境；
environment 环境变量；
transactionManager 事务管理器；
dataSource 数据源；
mappers 映射器；
```

2. 配置环境

```
<configuration>
    <environments default="development">
        <environment id="development">
            <transactionManager type="JDBC"/>
            <dataSource type="POOLED">
            <property name="driver" value="${driver}"/>
            <property name="url" value="${url}"/>
            <property name="username" value="${username}"/>
            <property name="password" value="${password}"/>
            </dataSource>
        </environment>
        <environment id="development2">
            ......
        </environment>
    </environments>
</configuration>
```

3. 基础配置文件——事务管理

MyBatis 有两种事务管理类型：

(1) JDBC:这个类型直接全部使用 JDBC 的提交和回滚功能,它依靠使用连接的数据源来管理事务的作用域。

(2) MANAGED:该类型无需做任何事，它从不提交 、回滚和关闭连接。而是让窗口来管理事务的全部生命周期。(比如说 Spring 或者 JavaEE 服务器)

数据源类型有三种：UNPOOLED、POOLED、JNDI。

① UNPOOLED:这个数据源实现只是在每次请求的时候简单地打开和关闭一个连接。虽然有点慢,但作为一些不需要性能和立即响应的简单应用来说，不失为一种好选择。

② POOLED:这个数据源缓存 JDBC 连接对象用于避免每次都要连接和生成连接实例而需要的验证时间。对于并发 WEB 应用,这种方式非常流行,因为它有最快的响应时间。

③ JNDI:这个数据源实现是为了准备和 Spring 或应用服务一起使用,可以在外部也可以在内部配置这个数据源,然后在 JNDI 上下文中引用它。这个数据源配置只需要两项属性：

4. 基础配置文件——SQL 映射文件

SQL 映射文件代码如下：

(1) 使用相对路径

```
<mappers>
    <mapper resource="org/mybatis/builder/UserMapper.xml"/>
    <mapper resource="org/mybatis/builder/AuthorMapper.xml"/>
    <mapper resource="org/mybatis/builder/BlogMapper.xml"/>
    <mapper resource="org/mybatis/builder/PostMapper.xml"/>
</mappers>
```

(2) 使用全路径

```
<mappers>
    <mapper url="file:///var/sqlmaps/AuthorMapper.xml"/>
    <mapper url="file:///var/sqlmaps/BlogMapper.xml"/>
    <mapper url="file:///var/sqlmaps/PostMapper.xml"/>
</mappers>
```

5. SqlSession 接口

(1) SqlSession 的获取方式

Reader reader = Resources.getResourceAsReader("configuration.xml"); SqlSessionFactory sqlSessionFactory = new SqlSessionFactoryBuilder().build(reader);

SqlSession sqlSession = sqlSessionFactory.openSession();

(2) SqlSession 的使用

调用 insert,update,selectList,selectOne,delete 等方法执行增删改查等操作。

4.8.5　应用示例

想要详细了解 MyBatis 推荐看一下官方的中文文档,文档对 MyBatis 介绍的很详细,在本章中不对 MyBatis 做很详细的介绍,只做个简单例子,在本例中调用 select 方法进行查询。

1. 准备工作

准备工作很简单,就两个 jar 包,一个是 MyBatis 的 jar 包;另一个是数据库的驱动。这里采用的是 mysql 数据库,具体 jar 包如图 4-19 所示。

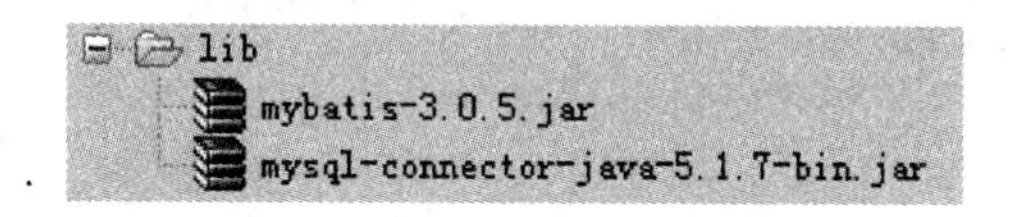

图 4-19　MyBatis 包文件

2. 数据库

使用的 test 库的 user 表,建表脚本如下:

```
drop table if exist user;
create table user(
    id int,
    name varchar(50)
);
insert user values(1,'张三');
```

```
insert user values(2,'李四');
insert user values(3,'王五');
```

3. 项目结构

结构如图 4－20 所示。

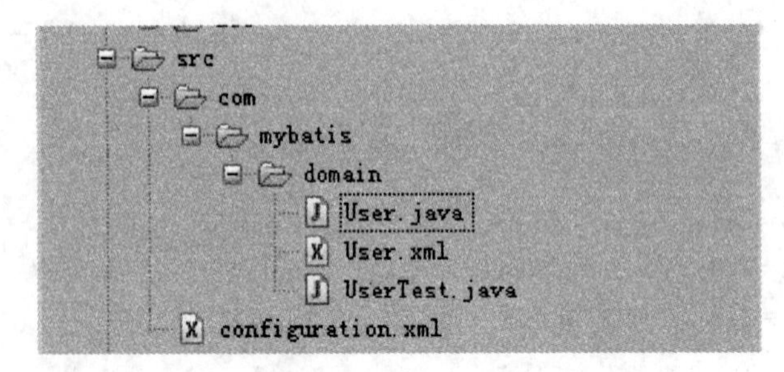

图 4－20　项目结构

4. 具体代码

User 类：

```
package com.mybatis.domain;
impor tjava.io.Serializable;
public class User implements Serializable{
    private static final long serialVersionUID = 1L;
    private int id;
    private String name;

    public User() {

    }

    public User(int id, String name) {
        this.id = id;
        this.name = name;
    }

    public int getId() {
        return id;
    }

    public void setId(int id) {
        this.id = id;
    }

    public String getName() {
        return name;
    }
```

```
    public void setName(String name) {
        this.name = name;
    }

    public String toString() {
        return "User [id:" + this.id + ";name:" + this.name + "]";
    }
}
```

5. 配置文件 configuration.xml

```
<?xml version="1.0" encoding="UTF-8" ?>
<!DOCTYPE configuration PUBLIC "-//mybatis.org//DTD Config 3.0//EN"
"http://mybatis.org/dtd/mybatis-3-config.dtd">
<configuration>
    <environments default="development">
        <environment id="development">
            <transactionManager type="JDBC" />
            <dataSource type="POOLED">
                <property name="driver" value="com.mysql.jdbc.Driver" />
                <property name="url"
    value="jdbc:mysql://localhost:3306/test? useUnicode=true&amp;characterEncoding=UTF
-8" />
                <property name="username" value="root" />
                <property name="password" value="root" />
            </dataSource>
        </environment>
    </environments>
    <mappers>
        <mapper resource="com/mybatis/domain/User.xml" />
    </mappers>
</configuration>
```

6. Sql 映射文件

```
<?xml version="1.0" encoding="UTF-8"?>
<!DOCTYPE mapper PUBLIC "-//mybatis.org//DTD Mapper 3.0//EN"
"http://mybatis.org/dtd/mybatis-3-mapper.dtd">

<mapper namespace="User">

    <cache />

    <select id="selectUser" parameterType="int"
        resultType="com.mybatis.domain.User">
```

```
            select * from user where id=#{id}
        </select>
</mapper>
```

7. 测试类

```
package com.mybatis.domain;
import java.io.IOException;
import java.io.Reader;
import org.apache.ibatis.io.Resources;
import org.apache.ibatis.session.SqlSession;
import org.apache.ibatis.session.SqlSessionFactory;
import org.apache.ibatis.session.SqlSessionFactoryBuilder;

public class UserTest {

    public static void main(String[] args) throws IOException {
        // TODO Auto-generated method stub
        String resource = "configuration.xml";
        Reader reader = Resources.getResourceAsReader(resource);
        SqlSessionFactory ssf = new SqlSessionFactoryBuilder().build(reader);
        SqlSession session = ssf.openSession();
        try {
            User user = (User) session.selectOne("selectUser","1");
            System.out.println(user);
        } catch (RuntimeException e) {
            // TODO Auto-generated catch block
            e.printStackTrace();
        }finally{
            session.close();
        }
    }

}
```

userTest.class 执行效果如图 4-21 所示。

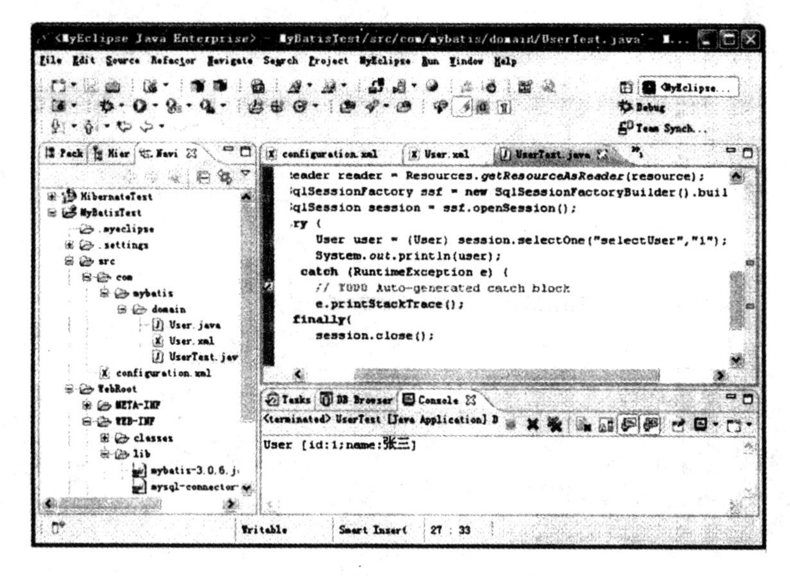

图 4-21　UserTest. clsss 执行效果

巩固练习

1. 描述 Hibernate 和 MyBatis 的工作机制。
2. 了解 Hibernate 和 MyBatis 两者的区别。
3. 进一步上机实践,完善书上例子。

第 5 章　Spring 应用

学习目标

1. 理解 Spring 框架运行机制。
2. 理解依赖注入的思想。
3. 使用 Spring 编程。

§5.1　Spring 概述

Spring 框架为基于 java 的企业应用程序提供了一个全面的编程和配置模型，它可以应用于任何类型的部署平台。Spring 是一个从实际开发中提炼出的框架，它完成了大量开发中的通用步骤。通过应用 Spring 框架，开发团队只需关注应用程序的业务逻辑，而各个应用层之间的通信管理交由框架处理。

Spring 框架的主要优势之一是分层架构，分层架构允许选择使用任一个组件，同时为 Java EE 应用程序开发提供集成的框架。Spring 框架的分层架构是由 7 个定义的模块组成。Spring 模块构建在核心容器之上，核心容器定义了创建、配置和管理 Bean 的方式，如图 5 - 1 所示。

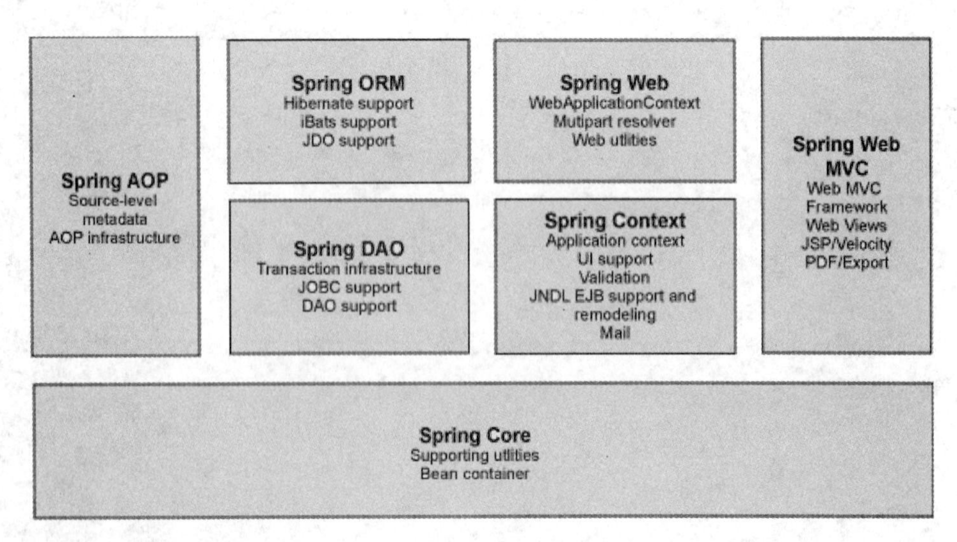

图 5 - 1　Spring 框架

组成 Spring 框架的每个模块(或组件)都可以单独存在,或者与其他一个或多个模块联合实现。各模块的功能如下:

① 核心容器。提供 Spring 框架的基本功能,其主要组件是 BeanFactory,是工厂模式的实现。

② Spring 上下文。向 Spring 框架提供上下文信息,包括企业服务,如 JNDI、EJB、电子邮件、国际化、校验和调度等。

③ Spring AOP。通过配置管理特性,可以很容易的使 Spring 框架管理的任何对象支持 AOP。Spring AOP 模块直接将面向方面编程的功能集成到 Spring 框架中。它为基于 Spring 应用程序的对象提供了事务管理服务。

④ Spring DAO。JDBC DAO 抽象层提供了有用的异常层次结构,用来管理异常处理和不同数据库供应商抛出的错误消息。异常层次结构简化了错误处理,并且极大地降低了需要编写的异常代码数量(如打开和关闭连接)。

⑤ Spring ORM。Spring 框架插入了若干 ORM 框架,提供 ORM 的对象关系工具,其中包括 JDO、Hibernate 和 iBatis SQL Map,并且都遵从 Spring 的通用事务和 DAO 异常层次结构。

⑥ Spring Web 模块。为基于 Web 的应用程序提供上下文。它建立在应用程序上下文模块之上,简化了处理多份请求及将请求参数绑定到域对象的工作。Spring 框架支持与 Jakarta Struts 的集成。

⑦ Spring MVC 框架,是一个全功能构建 Web 应用程序的 MVC 实现。通过策略接口实现高度可配置,MVC 容纳了大量视图技术,其中包括 JSP、Velocity、Tiles、iText 和 POI。

Spring 的核心机制是依赖注入,也称为控制反转。在介绍它之前,首先介绍一种设计模式——简单工厂模式。

§5.2　简单工厂模式

工厂模式是通过专门定义一个类来负责创建其他类的实例,被创建的实例通常都具有共同的父类。如果简单工厂模式所涉及到的具体产品之间没有共同的逻辑,那么就可以使用接口来扮演抽象产品的角色。通过该模式一定程度上提高了程序的复用性,降低了代码的耦合度。

下面举例说明工厂模式的应用。

创建一个 Java Project,命名为“FactoryExample”。在 src 文件夹下建立包 face,在该包下建立接口 Animal,代码如下:

```
package face;
public interface Animal
 {
      void eat();
      void walk();
 }
```

在 src 文件夹下建立包 iface，在该包下建立 Cat 类和 Dog 类，分别实现 Animal 接口。Cat. java 代码如下：

```
package iface;
import face. Animal;
public class Cat implements Animal{
    public void eat()
{
        System. out. println("猫喜欢吃鱼!");
    }
    public void walk() {
        System. out. println("猫爬树敏捷!");
    }
}
```

创建 Dog. java 文件，代码如下：

```
package iface;
import face. Animal;
public class Dog implements Animal
{
    public void eat() {
        System. out. println("狗喜欢啃骨头!");
    }
    public void walk() {
        System. out. println("狗经常跑!");
    }
}
```

在 src 文件夹下建包 factory，在该包内建立工厂类 Factory，代码如下：

```
package Factory;
import iface. Cat;
import iface. Dog;
import face. Animal;
public class MyFactoy {
    public Animal getAnimal(String name){
        if(name. equals("Cat")){
            return new Cat();
        }else if(name. equals("Dog")){
            return new Dog();
```

```
        }else{
            throw new IllegalArgumentException("参数不正确");
        }
    }
}
```

在 src 文件夹下建包 test，在该包内建立测试类 Test，代码如下：

```
package test;
import face.Animal;
import Factory.MyFactoy;
public class test
{
    public static void main(String[] args)
    {
        Animal animal=null;
        animal=new MyFactoy().getAnimal("Cat");
        animal.eat();
        animal.walk();
        animal=new MyFactoy().getAnimal("Dog");
        animal.eat();
        animal.walk();
    }
}
```

该程序为 Java 应用程序，直接运行可看出结果，如图 5-2 所示。

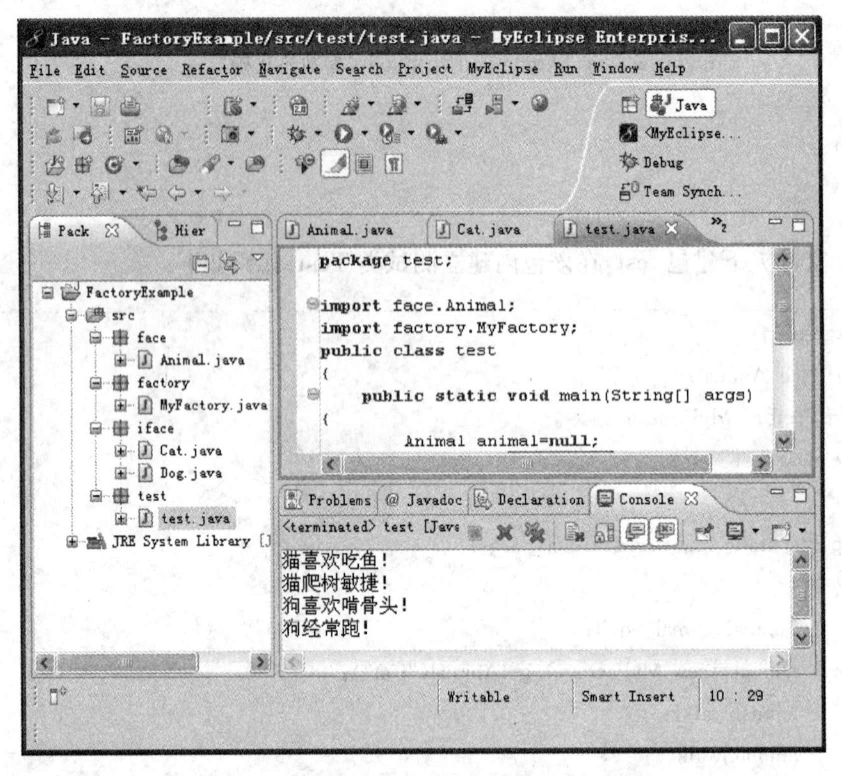

图 5-2 工厂模式运行结果

在简单工厂模式中，调用程序无需直接创建所调用类的实例，都是通过工厂类实现实例化，从而降低了程序间的耦合度。而 Spring 框架则提供了更好的办法，开发人员可以直接应用 Spring 提供的依赖注入方式，即被调用者的实例工作由 Spring 容器完成，让 bean 与 bean 之间以配置文件组织在一起，对象间的具体实现互相透明，即降低了程序间耦合度，又减轻了开发者的负担。下面用具体实例来介绍依赖注入的使用方法。

§5.3 依赖注入应用

1. 为项目添加 Spring 开发能力

右击项目名，选择【MyEclipse】→【Add Spring Capabilities…】菜单项，将出现如图 5-3 所示的对话框，选中要应用的 Spring 的版本及所需的类库文件。

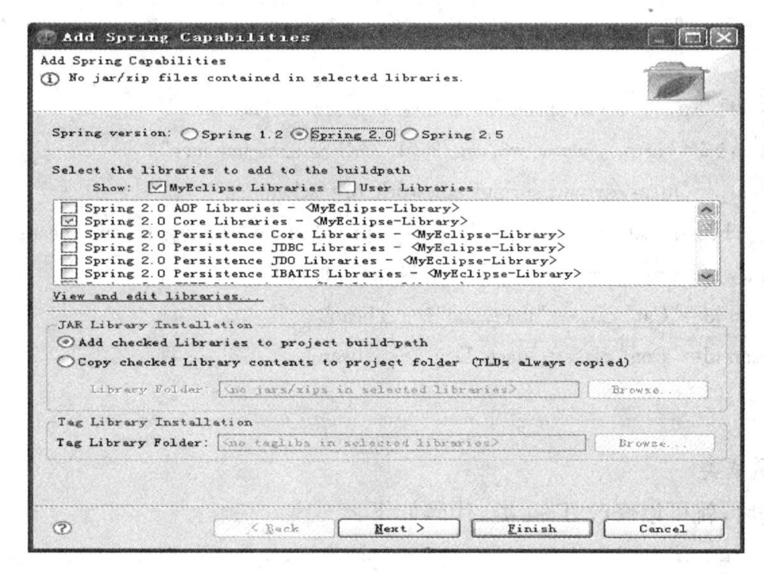

图 5-3　添加 Spring 运行所需类

选择结束后，单击【Next】按钮，出现如图 5-4 所示的界面。用于创建 Spring 的配置文件。

单击【Finish】按钮完成。项目的 src 文件夹下会出现名为 applicationContext. xml 的文件，这就是 Spring 的核心配置文件。

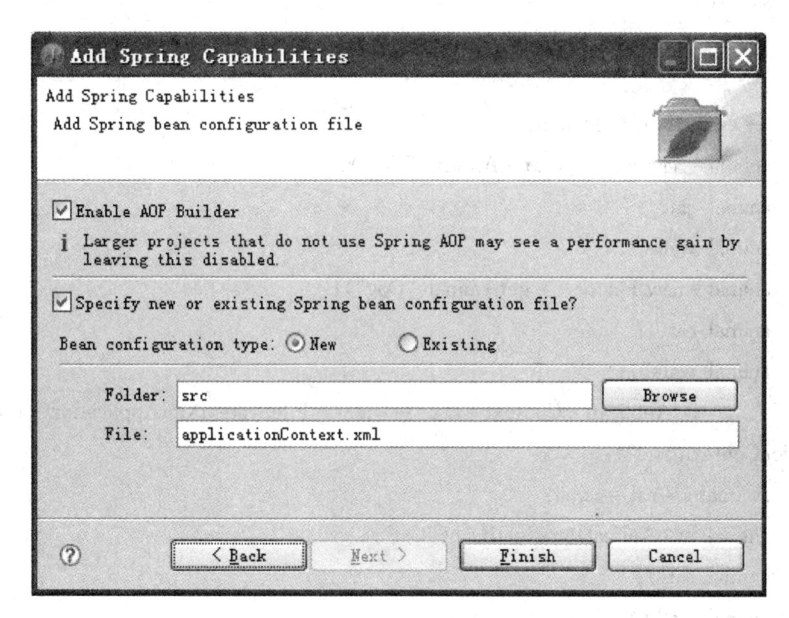

图 5-4　生成 applicationContext. xml

2. 修改配置文件 applicationContext. xml

修改后，代码如下：

```
<?xml version="1.0" encoding="UTF-8"?>
<beans
    xmlns="http://www.springframework.org/schema/beans"
    xmlns:xsi="http://www.w3.org/2001/XMLSchema-instance"
    xmlns:p="http://www.springframework.org/schema/p"
    xsi:schemaLocation="http://www.springframework.org/schema/beans http://www.springframework.org/schema/beans/spring-beans-3.0.xsd">
    <bean id="Cat" class="iface.Cat"></bean>
    <bean id="Dog" class="iface.Dog"></bean>
</beans>
```

3. 修改测试类

配置完成后，就可以修改 Test 类，代码如下：

```
package test;
import org.springframework.context.ApplicationContext;
import org.springframework.context.support.FileSystemXmlApplicationContext;
import face.Animal;
import Factory.MyFactoy;
public class test {
    public static void main(String[] args)
        {
        /* Animal animal=null;
        animal=new Factory().getAnimal("Cat");
        animal.eat();
        animal.walk();
        animal=new Factory().getAnimal("Dog");
        animal.eat();
        animal.walk();
            */ApplicationContext ctx = new FileSystemXmlApplicationContext("src/applicationContext.xml");
        Animal animal = null;
        animal = (Animal) ctx.getBean("Cat");
        animal.eat();
        animal.walk();
        animal = (Animal) ctx.getBean("Dog");
        animal.eat();
        animal.walk();
    }
}
```

4. 运行

运行该测试类,结果如图5-5所示。

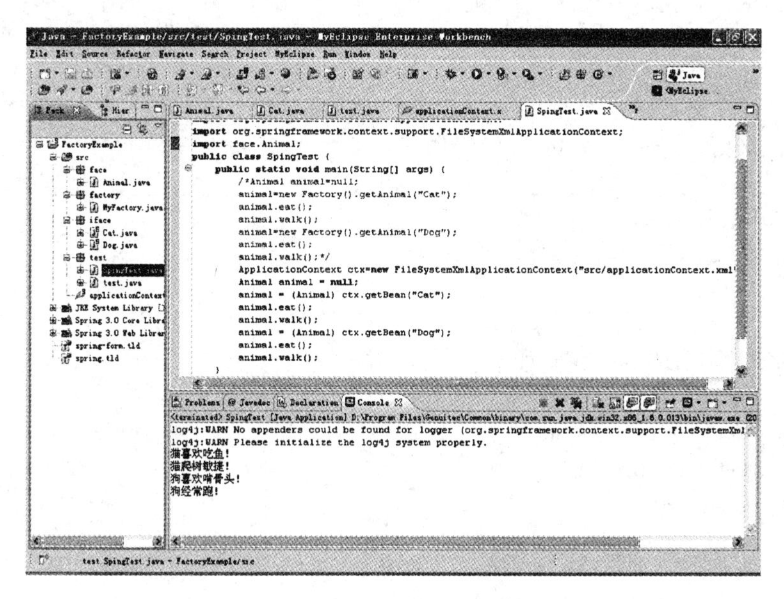

图5-5 Spring依赖注入运行结果

对象ctx就相当于原来的Factory工厂,原来的Factory可以删除掉了。再回头看原来的applicationContext.xml文件配置:

```
<bean id="cat" class="iface.Cat"></bean>
<bean id="dog" class="iface.Dog"></bean>
```

id是ctx.getBean的参数值,一个字符串。class是一个类(包名+类名)。然后在Test类里获得Chinese对象及American对象:

```
animal = (Animal) ctx.getBean("cat");
animal = (Animal) ctx.getBean("dog");
```

§5.4 Spring注入方式

依赖注入通常有如下两种。

设值注入:IoC容器使用属性的setter方法来注入被依赖的实例。

构造注入:IoC容器使用构造器来注入被依赖的实例。

5.4.1 设值注入

设值注入是通过setter方法注入被调用者的实例。这种方法简单、直观,很容易理解,因而在Spring的依赖注入被经常使用,下面举例说明。

创建一个Java Project,命名为“FactoryExample1”。在项目的src文件夹下建立下面的源文件。

Animal 的接口，Animal.java，代码如下：

```
public interface Animal {
    void sport();
}
```

Behaviour 接口，Behaviour.java，代码如下：

```
public interface Behaviour {
    public String kind();
}
```

Animal 实现类 Cat.java，代码如下：

```
public class Cat implements Animal {
    private Behaviour beh;
    public void sport() {
        System.out.println(beh.kind());
    }
    public void setBeh(Behaviour beh) {
        this.beh = beh;
    }
}
```

Behaviour 实现类 Climb.java，代码如下：

```
public class Climb implements Behaviour{
    public String kind() {
        return "猫也会爬!";
    }
}
```

通过 Spring 的配置文件来完成其对象的注入，代码如下：

```
<?xml version="1.0" encoding="UTF-8"?>
<beans
    xmlns="http://www.springframework.org/schema/beans"
    xmlns:xsi="http://www.w3.org/2001/XMLSchema-instance"
    xsi:schemaLocation="http://www.springframework.org/schema/beans
    http://www.springframework.org/schema/beans/spring-beans-2.0.xsd">
    <!-- 定义第一个 Bean，注入 Cat 类对象 -->
    <bean id="cat" class="Cat">
        <!-- property 元素用来指定需要容器注入的属性，lan 属性需要容器注入
                ref 就指向 lan 注入的 id -->
        <property name="beh" ref="climb "></property>
    </bean>
```

```
    <!-- 注入 Climb -->
    <bean id=" climb " class=" Climb "></bean>
</beans>
```

测试代码如下：

```
import org.springframework.context.ApplicationContext;
import org.springframework.context.support.FileSystemXmlApplicationContext;
public class Test {
    public static void main(String[] args) {
        ApplicationContext ctx = new FileSystemXmlApplicationContext("src/applicationContext.
xml");
        Animal animal = null;
        animal = (Animal) ctx.getBean("cat");
        animal.sport();
    }
}
```

程序执行结果如图 5-6 所示。

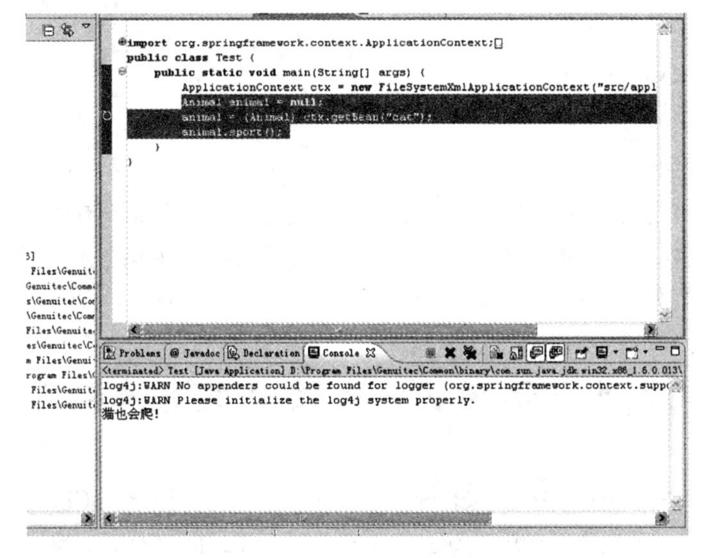

图 5-6　Spring 设置注入运行结果

5.4.2　构造注入

只需对前面的 Cat 类做简单的修改：

```
public class Cat implements Animal{
    private Behaviour beh;
    public Cat(){
    };
    // 构造注入所需要的带参数的构造函数
```

```
    public Cat(Behaviour beh){
        this. beh=beh;
    }
    public void sport() {
        System. out. println(beh. kind());
    }
}
```

配置文件需要做简单的修改：

```
<?xml version="1. 0" encoding="UTF-8"?>
<beans
    xmlns="http://www. springframework. org/schema/beans"
    xmlns:xsi="http://www. w3. org/2001/XMLSchema-instance"
    xsi:schemaLocation="http://www. springframework. org/schema/beans
    http://www. springframework. org/schema/beans/spring-beans-3. 0. xsd">
    <!-- 定义第一个 Bean,注入 Cat 类对象 -->
    <bean id="cat" class="Cat">
    <!-- 使用构造注入,为 Cat 实例注入 Behaviour 实例 -->
          <constructor-arg ref="climb"></constructor-arg>
    </bean>
    <!-- 注入 Climb -->
    <bean id=" climb " class=" Climb "></bean>
</beans>
```

在开发工厂中,这两种注入方式都是常用的。两种注入的方式没有绝对的好坏,只是适应场景有所不同。建议采用设值注入为主,构造注入为辅的注入策略。对于依赖关系无需变化的注入,尽量采用构造注入。而其他的依赖关系的注入,则考虑采用设值注入。

§5.5 Spring 核心接口及基本配置

Spring 有两个核心接口:BeanFactory(Bean 工厂)和 ApplicationContext(应用上下文),其中 ApplicationContext 是 BeanFactory 的子接口。它们都可以代表 Spring 容器。Bean 是 Spring 管理的基本单元,Spring 容器是生成 Bean 实例的工厂,并管理容器中的 Bean。在基于 Spring 的 Java EE 应用中,所有的组件都被当成 Bean 处理,包括数据源、Hibernate 的 SessionFactory、事务管理器等。

5.5.1 Spring 核心接口

1. BeanFactory

Spring 容器最基本的接口就是 BeanFactory。在 Spring 中有几种 BeanFactory 的实现,其中最常用的是 org. springframework. bean. factory. xml. XmlBeanFactory。它根据 XML 文件中的定义装载 Bean。

要创建 XmlBeanFactory,需要传递一个 java. io. InputStream 对象给构造函数。InputStream

对象提供 XML 文件给工厂。例如，下面的代码片段使用一个 java. io. FileInputStream 对象把 Bean XML 定义文件给 XmlBeanFactory：

```
BeanFactory factory = new XmlBeanFactory(new FileInputStream(" applicationContext. xml"));
```

通过代码告诉 Bean Factory，从 XML 文件中读取 Bean 的定义信息，然而现在 Bean Factory 没有实例化 Bean，Bean 被延迟载入到 Bean Factory 中，就是说 Bean Factory 会立即把 Bean 定义信息载入进来，但是 Bean 只有在需要的时候才会被实例化。

为了从 BeanFactory 得到 Bean，只要简单地调用 getBean()方法，把需要的 Bean 的名字当做参数传递进去就行了。由于得到的是 Object 类型，所以要进行强制类型转化。

```
MyBean myBean = (MyBean)factory. getBean("myBean");
```

2. ApplicationContext

两者都是载入 Bean 定义信息，装配 Bean，根据需要分发 Bean。但是 ApplicationContext 提供了一些附加功能：

① 应用上下文提供了文本信息解析工具，包括对国际化的支持。

② 应用上下文提供了载入文本资源的通用方法，如载入图片。

③ 应用上下文可以向注册为监听器的 Bean 发送事件。

由于它提供的附加功能，应用系统选择 ApplicationContext 作为 Spring 容器更方便些。在 ApplicationContext 的诸多实现中，有三个常用的实现：

① ClassPathXmlApplicationContext：从类路径中的 XML 文件载入上下文定义信息，把上下文定义文件当成类路径资源。

② FileSystemXmlApplicationContext：从文件系统中的 XML 文件载入上下文定义信息。

③ XmlWebApplicationContext：从 Web 系统中的 XML 文件载入上下文定义信息。

例如：

```
ApplicationContext context=new FileSystemXmlApplicationContext ("c:/foo. xml");
ApplicationContext context=new ClassPathApplicationContext ("foo. xml");
ApplicationContext context = WebApplicationContextUtils. getWebApplicationContext (request. getSession
(). getServletContext ());
```

FileSystemXmlApplicationContext 和 ClassPathXmlApplicationContext 的区别是：FileSystemXmlApplicationContext 只能在指定的路径中寻找 foo. xml 文件，而 ClassPathXml ApplicationContext 可以在整个类路径中寻找 foo. xml 文件。

5. 5. 2　Spring 基本配置

理论上，Bean 装配可以从任何配置资源获得。但实际上，XML 是最常见的 Spring 应用系统配置源。

如下的 XML 文件展示了一个简单的 Spring 上下文定义文件：

```
<?xml version="1. 0" encoding="UTF-8"?>
...
<beans ...>                                   // 根元素
```

```
    <bean id="cat" class="iface.Cat"/>        // Bean 实例
    <bean id="dog" class="iface.Dog"/>          // Bean 实例
</beans>
```

在 XML 文件定义 Bean，上下文定义文件的根元素<beans>。<beans>有多个<bean>子元素。每个<bean>元素定义了一个 Bean(任何一个 Java 对象)如何被装配到 Spring 容器中。

5.5.3　Spring 容器中的 Bean

向 Spring 容器中添加一个 Bean 只需要向 XML 文件中添加一个<bean>元素。如下面的语句：

```
<bean id="cat" class="iface.Cat"/>
```

当通过 Spring 容器创建一个 Bean 实例时，不仅可以完成 Bean 实例的实例化，还可以为 Bean 指定特定的作用域。

① 原型模式与单实例模式：Spring 中的 Bean 默认情况下是单实例模式。在容器分配 Bean 的时候，它总是返回同一个实例。

<bean>的 singleton 属性告诉 ApplicationContext 这个 Bean 是不是单实例 Bean，默认是 true，但是把它设置为 false 的话，就把这个 Bean 定义成了原型 Bean。

```
<bean id="cat" class="iface.Cat" singleton="false"/>     //原型模式 Bean
```

② request 或 session：对于每次 HTTP 请求或 HttpSession，使用 request 或 session 定义的 Bean 都将产生一个新实例，即每次 HTTP 请求或 HttpSession 将会产生不同的 Bean 实例。

③ global session：每个全局的 HttpSession 对应一个 Bean 实例。典型情况下，仅在使用 portlet context 的时候有效。

当一个 Bean 实例化的时候，有时需要做一些初始化的工作，然后才能使用。因此，Spring 可以在创建和拆卸 Bean 的时候调用 Bean 的两个生命周期方法。

在 Bean 的定义中设置自己的 init-method，这个方法在 Bean 被实例化时马上被调用。同样，也可以设置自己的 destroy-method，这个方法在 Bean 从容器中删除之前调用。

一个典型的例子是连接池 Bean，具体代码如下：

```
public class MyConnectionPool{
    ...
    public void initalize(){
    //initialize connection pool}
    public void close(){
    //release connection}
    ...
}
```

Bean 的定义如下：

```
<bean id="connectionPool" class="com.spring.MyConnectionPool"
      init-method="initialize"       //当 Bean 被载入容器时调用 initialize 方法
      destroy-method="close">    //当 Bean 从容器中删除时调用 close 方法
</bean>
```

§5.6　AOP

5.6.1　AOP 简介

AOP(Aspect Orient Programming),也就是面向切面编程,作为面向对象编程的一种补充。AOP 与 OOP 互为补充,可以这样理解,面向对象编程是从静态角度考虑程序结构,面向切面编程是从动态角度考虑程序运行过程。AOP 专门用于处理系统中分布于各个模块中的交叉关注点的问题,在 Java EE 应用中,通过 AOP 来处理一些具有横切性质的系统级服务,如事务管理、安全检查、缓存、对象池管理等。AOP 的核心就是动态代理,接下来将使用用户登录程序案例来理解代理机制。

5.6.2　代理机制

程序中经常需要在其中写入与本功能不是直接相关但很有必要的代码,如日志记录,信息发送,安全和事务支持等,以下代码是一个用户注册类的代码:

```
package com.sitinspring.login;

/**
 * 用于用户登录类
 */
public class LoginService{
    public void register(String name,String pswd,String email){
        Logger.log("用户将登录"+name);
        System.out.println("存储用户信息");
        MailSender.send(email, "欢迎"+name+"用户登录本系统");
    }

    public static void main(String[] args){
        // 调用示例
        LoginService service=new LoginService();
        service.register("sitinspring","123456","wang@njxzc.edu");
    }
}
```

Logger.java 代码如下:

```
package com.sitinspring.newlogin;

import java.text.Format;
import java.text.SimpleDateFormat;
import java.util.Date;
/**
 * 模拟记录器
 */
public class Logger{
    public static void log(String str){
        System.out.println(getCurrTime()+"INFO:"+str);
    }

    /**
     * 取得当前时间
     * @return
     */
    private static String getCurrTime() {
        Date date = new Date();
        Format formatter = new SimpleDateFormat("HH时mm分ss秒");
        return formatter.format(date);
    }
}
```

MailSender.java 代码如下：

```
package com.sitinspring.newlogin;
/**
 * 模拟邮件发送器
 */
public class MailSender{
    public static void send(String email,String concept){
        System.out.println("向"+email+"发送邮件内容为:"+concept+"的邮件");
    }
}
```

从 LoginService.java 文件代码可以看出，日志和信息发送等操作并不属于 LoginService 逻辑，这增加了程序的耦合度，并且是程序逻辑不清。如果一旦不需要日志或信息发送等服务，将要修改所有与日志记录和信息发送动作有关的代码，给程序维护带来不便。

这种情况可以使用代理(Proxy)机制来解决，代理可以提供对另一个对象的访问，同时隐藏实际对象的具体细节，一般会实现它所表示的实际对象的接口；代理可以访问实际对象，但是延迟实现实际对象的部分功能，实际对象实现系统的实际功能，代理对象对客户隐藏了实际对象。客户不知道它是与代理打交道还是与实际对象打交道。

若使用代理模式，可把枝节性代码放入代理类中，这样主干性代码保持在真实的类中，这样就能有效降低耦合度。通过在耦合紧密的类之间引入一个中间类是降低类之间的耦合度的是最常用做法。

具体来说就是把枝节性代码放入代理类中，它们由代理类负责调用，而真实类只负责主干的核心业务，它也由代理类调用，它并不知道枝节性代码的存在和作用。对外来说，代理类隐藏在接口之后，客户并不清楚也不需要清楚具体的调用过程。通过这样的处理，主干与枝节之间的交叉解开了，外界的调用也没有复杂化，这就有效降低系统各部分间的耦合度。代理有两种代理方式：静态代理（static proxy）和动态代理（dynamic proxy）。

1. 静态代理

在静态代理的实现中，代理类与被代理的类必须实现同一个接口。在代理类中可以实现日志记录或信息发送等相关服务，并在需要的时候再呼叫被代理类。这样被代理类就可以仅仅保留业务相关的职责。下面就来了解一下静态代理的方法实现代码。

首先定义有关 IService 接口，IService. java 代码如下：

```
package com. sitinspring. newlogin;
/ * *
  * Service 接口
  * /
public interface IService {
    public void login(String name,String pswd,String email);
}
```

然后让实现业务逻辑的 LoginService 类实现 IService 接口，LoginService. java 代码如下：

```
package com. sitinspring. newlogin;
/ * *
  * 用于用户登录类
  * /
public class LoginService implements IService{
    public void register(String name,String pswd,String email){
        // 真正需要由本函数担负的出理
        System. out. println("存储用户信息");
    }
}
```

可以看到，在 LoginService 类中没有任何日志或信息发送的代码插入其中，日志和信息发送服务的实现将被放到代理类中，代理类同样要实现 IHello 接口。

LoginProxy. java 代码如下：

```
package com. sitinspring. newlogin;
import com. sitinspring. login. Logger;
import com. sitinspring. login. MailSender;
public class LoginProxy implements IService{
    private IService serviceObject;
    public LoginProxy(IService serviceObject){
        this. serviceObject=serviceObject;
    }
```

```
    @Override
    public void login(String name, String pswd, String email) {
        Logger.log("用户将登录"+name);
        serviceObject.login(name, pswd, email);;
        MailSender.send(email, "欢迎"+name+"用户登录本系统");
    }
}
```

在 LoginProxy 类的 login()方法中,真正实现业务逻辑前后安排记录服务,可以实际撰写一个测试程序来看看如何使用代理类。

ProxyDemo.java 代码如下:

```
package com.sitinspring.newlogin;

public class ProxyDemo {

    public static void main(String[] args) {
        // TODO Auto-generated method stub
        IService proxy=new LoginProxy(new LoginService());
        proxy.login("njxzc","123456", "wang@njxzc.edu");
    }
}
```

程序运行结果如图 5-7 所示。

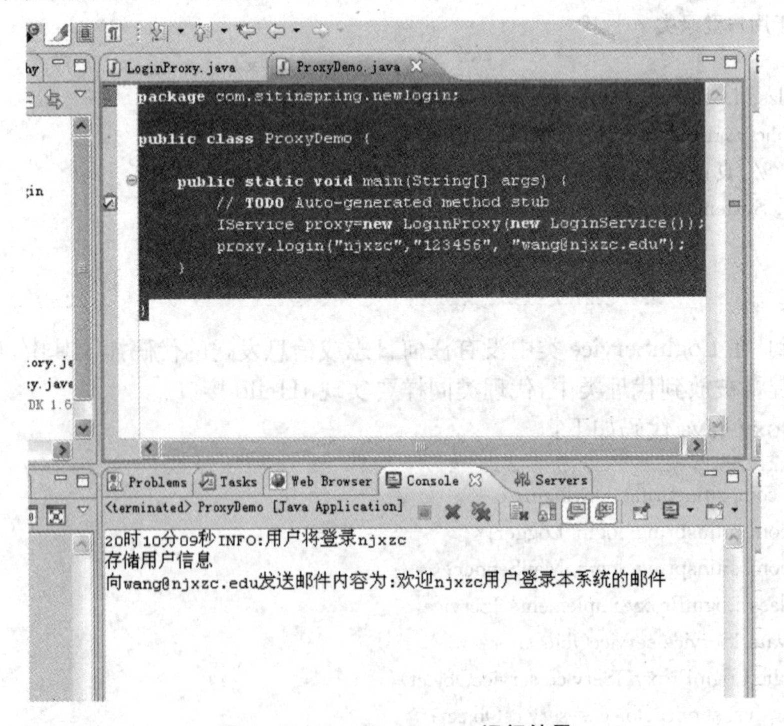

图 5-7　ProxyDemo.java 运行结果

2. 动态代理

使用动态代理可以使得一个处理者(Handler)为各个类服务。要实现动态代理,同样需要定义所要代理的接口。

首先定义有关 IService 接口,IService. java 代码如下:

```
package com. sitinspring. newlogin;
/ * *
  * Service 接口
  * /
public interface IService {
    public void login(String name,String pswd,String email);
}
```

然后让实现业务逻辑的 LoginService 类实现 IService 接口,LoginService. java 代码如下:

```
package com. sitinspring. newlogin;
/ * *
  * 用于用户登录类
  * /
public class LoginService implements IService{
    public void register(String name,String pswd,String email){
        // 真正需要由本函数担负的出理
        System. out. println("存储用户信息");
    }
}
```

以上两个文件跟静态代理是一样的,但下面的代理类是不同的。

LoginServiceProxy. java 代码如下:

```
package com. sitinspring. newlogin;
import java. lang. reflect. InvocationHandler;
import java. lang. reflect. InvocationTargetException;
import java. lang. reflect. Method;
import com. sitinspring. login. Logger;
import com. sitinspring. login. MailSender;

/ * *
  * 注册服务代理类
  * @author: sitinspring(junglesong@gmail. com)
  * @date: 2008 - 5 - 27 -下午 09:45:10
  * /
public class LoginServiceProxy implements InvocationHandler {
    // 代理对象
    Object obj;
```

```
    //构造函数,传入代理对象
    public LoginServiceProxy(Object o) {
        obj = o;
    }
    /* *
     * 调用被代理对象的将要被执行的方法,我们可以在调用之前进行日志记录,之后执行邮件
发送
     */
    public Object invoke(Object proxy, Method m, Object[] args) throws Throwable {
        Object result = null;
        try {
            // 进行日志记录
            String name=(String)args[0];
            Logger.log(name+"用户将登录");
            //调用 Object 的方法
            result = m.invoke(obj, args);
             // 执行邮件发送
            String email=(String)args[2];
            MailSender.send(email, "欢迎"+name+"用户登录本系统");
        } catch (InvocationTargetException e) {
        } catch (Exception eBj) {
        } finally {
            // Do something after the method is called ...
        }
        return result;
    }
}
```

该代理类的内部属性为 Object 类,使用时通过该类的构造函数 LoginServiceProxy(Object obj)对其赋值;此外,在该类还实现了 invoke 方法,该方法中的 method.invoke(obj, args)

其实就是调用被代理对象的将要被执行的方法。这种实现方式是通过反射实现的,方法参数 obj 是实际的被代理对象,args 为执行被代理对象相应操作所需的参数。通过动态代理类,可在调用之前或调用之后执行一些相关操作。代理类的实例需要特殊的方式生成,LoginServiceFactory.java 代码如下:

```
package com.sitinspring.newlogin;
import java.lang.reflect.Proxy;


/* *
 * 工厂类,用于隐藏内部细节
 */
public class LoginServiceFactory{
    public static IService genereteService(){
```

```
        return (IService)Proxy.newProxyInstance(
                IService.class.getClassLoader(),
                new Class[]{IService.class},
                new LoginServiceProxy(new LoginService()));
    }
}
```

Proxy 即为 java 中的动态代理类，其方法 Static Object newProxyInstance(ClassLoader loader, Class[] interfaces, InvocationHandler h)：返回代理类的一个实例，其中 loader 是类加载器，interfaces 是被代理的真实类的接口，h 是具体的代理类实例。

动态代理是在运行时生成的类，在生成时必须提供一组接口给它，然后该类就宣称它实现了这些接口，可以把该类的实例当作这些接口中的任何一个实现类来用。这个动态代理类就是一个代理，它不会做实质性的工作，而是在生成它的实例时须提供一个真实的类的实例，由它接管实际的工作。

最后写一个测试程序，main.java 代码如下：

```
package com.sitinspring.newlogin;

public class Main{
    public static void main(String[] args){
        // 调用示例
        IService service=LoginServiceFactory.genereteService();
        service.login("njxzc","123456","wang@njxzc.edu");
    }
}
```

main.java 程序运行结果与静态代理方法一致。

Proxy 即为 java 中的动态代理类，其方法 Static Object newProxyInstance(ClassLoader loader, Class[] interfaces, InvocationHandler h)：返回代理类的一个实例，其中 loader 是类加载器，interfaces 是被代理的真实类的接口，h 是具体的代理类实例。

使用代理类将与业务逻辑无关的动作提取出来，设计为一个服务类，如同前面的范例 LoginServiceProxy，这样的类称为切面(Aspect)。AOP 将日志记录这类动作设计为通用，不介入特定业务类的一个职责清楚的 Aspect 类，这就是所谓的 Aspect-Oriented Programming。

巩固练习

1. 描述 Spring 依赖注入的工作机制
2. 进一步完善书中例子。

第 6 章　Struts 2、MyBatis 和 Spring 整合应用——教务管理系统开发实例

学习目标

1. 了解并掌握 Struts 2、MyBatis、Spring(SMI) 三种框架整合应用的配置。
2. 掌握项目开发流程。
3. 巩固所学框架知识。

§6.1　项目简介

该项目是由某校师生与企业联合开发出一个学院教务管理系统，通过该项目的锻炼，使师生对项目开发过程、项目管理和最新开发框架有了充分了解。本章通过对以上整体项目的简化，以教务系统中教学培养方案的管理功能作为实例，描述了几个框架综合应用的实现过程。

1. 系统主页面(如图 6-1 所示)

图 6-1　系统首页

2. 教务系统整体功能需求说明

(a) 院长功能模块

① 工作进度浏览及催办；

② 审核发布；

③ 统计信息；

④ 密码管理；

⑤ 院长信箱；

⑥ 待办事项。

(b) 教师功能模块

(1) 我的教学

① 教学任务；

② 教学文档；

③ 修改申请。

(2) 文档查询

(3) 个人管理

① 密码管理；

② 个人信息。

(4) 教师邮箱

(5) 待办事项

(c) 专业负责人功能模块

① 培养方案管理模块；

② 课程组/组员管理模块；

③ 信息统计模块；

④ 个人信息管理模块。

(d) 管理员功能模块

① 账号管理功能；

② 文档的过程统计和报表功能；

③ 历史查询功能；

④ 消息提醒功能；

⑤ 截止日期设置功能；

⑥ 个人信息管理功能。

(e) 学生功能模块

(f) 教研室主任功能模块

① 过程控制；

② 代办事务；

③ 信息统计模块；

④ 个人信息管理模块。

3. 开发模式

该项目采用敏捷迭代的开发模式，迭代开发计划模版如图 6－2 所示。

模块	功能	开发人员	迭代	开发开始日期	开发结束日期	review责任人	Story状态	测试人员
	我的教学-教学任务	李朔	P1					
	我的教学-教学文档	李朔	P1					
	我的教学-修改申请	李朔	P1					
	个人信息	李朔	P1					
	培养方案管理	王小正	P1					
	个人信息管理模块	王小正	P1					
	账号管理功能	侯青	P1					
	历史查询功能	侯青	P1					
	消息提醒功能	张俊	P1					
	截止日期 设置功能	沙先国	P1					
	个人信息管理功能	张颖	P1					
	登录	张颖	P1					
	角色管理	沙先国	P1					
	权限管理	沙先国	P1					
	教学计划管理	徐家喜	P1					
	任课老师管理	张俊	P1					
	教学任务管理	徐家喜	P1					

图 6-2　开发计划模板

开发过程中每个成员每个阶段按计划提交 Story，story 模版见附录 A。

4. 数据库设计

培养方案表的数据字典如表 6-1 所示。

表 6-1　教学大纲数据字典

字段名	字段类型	缺省值	中文显示名称	约束关系	意义
cultivationProgramID	int	自增长	培养方案 ID	非空	主键
cultivationTarget	text		培养目标内容		
cultivationSpecification	text		培养规格内容		
schoolingPeriod	VARCHAR(1)		学制内容		3 或 4 或 5
degreeID	VARCHAR(10)		学位内容		工学学士或理学学士
graduationRequirement	text		毕业具体条件		
creator	VARCHAR(20)		创建人	非空	培养方案制定人
createTime	DATE		创建时间		自动插入当前时间
schoolYear	DATE		入学年份		
professionCode	VARCHAR(2)		专业（方向）代码	非空	
cultivationProgramName	VARCHAR(50)		培养方案名称		

其余数据库字典见附录 B。

§6.2　技术架构

1. 平台开发环境

(1) Spring 3、MyBatis 3 和 Struts 2 作为技术构架；

(2) Tomcat 6.0；

(3) Mysql 5；

(4) JDK 6.X；

(5) Eclipse Java EE IDE for Web Developers；

2. 系统业务分层结构(如图 6－3 所示)

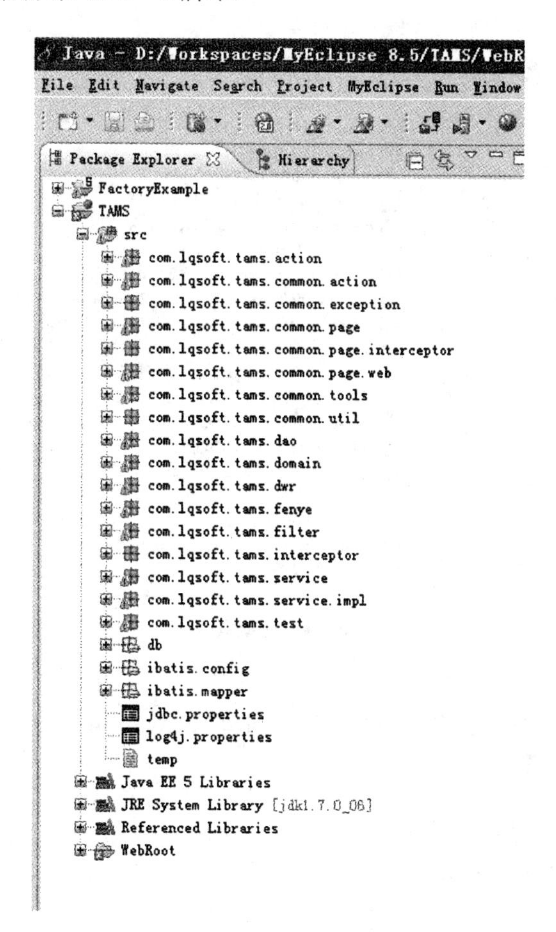

图 6－3　系统业务分层结构

主要步骤如下：

① 创建 Web Project；

② 创建包；

③ 修改“web.xml”，添加“struts.xml”文件；

④ 加载用户自定义包，修改“web.xml”，添加“struts.properties”文件；

⑤ 加载 MyBatis 框架；

⑥ 对数据库表进行反向工程，生成映射文件。

§6.3 项目创建流程（以创建培养方案为例）

步骤 1：

创建名为 njxzctms 的 Web Project，添加 Struts 2、Spring 及 MyBatis 库文件，创建包如图 6-4 所示。

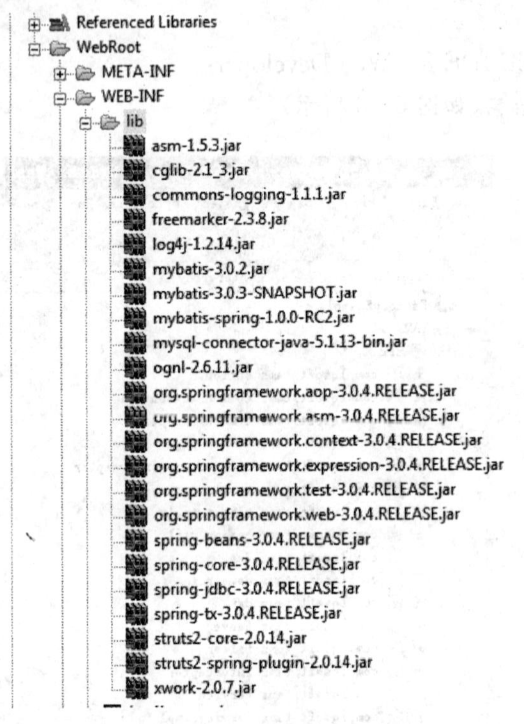

图 6-4 系统库文件

步骤 2：

在 Mysql 数据库中创建 JWXT 库，在库中创建 cultivationprogram（培养方案）表，创建表的 sql 语句如下。

```
-- ----------------------------------------
-- Table structure for 'cultivationprogram'
-- ----------------------------------------
DROP TABLE IF EXISTS 'cultivationprogram';
CREATE TABLE 'cultivationprogram' (
    'cultivationProgramID' int(11) NOT NULL AUTO_INCREMENT,
    'cultivationTarget' text,
    'cultivationSpecification' text,
    'schoolingPeriod' varchar(1) DEFAULT NULL,
```

```
  'degreeName' varchar(10) DEFAULT NULL,
   'graduationRequirement' text,
   'creator' varchar(20) DEFAULT NULL,
   'createTime' date DEFAULT NULL,
   'schoolYear' int(4) DEFAULT NULL,
   'professionCode' varchar(2) DEFAULT NULL,
   'cultivationName' varchar(50) DEFAULT NULL,
PRIMARY KEY (' cultivationProgramID'),
   UNIQUE KEY 'cultivationName' (' cultivationName')
) ENGINE=InnoDB AUTO_INCREMENT=60 DEFAULT CHARSET=utf8;
```

步骤 3：创建 mysql. properties 文件，内容如下。

```
## JDBC  Mysql propertiesa
```

dbc. driverClassName=com. mysql. jdbc. Driver

jdbc. url = jdbc \: mysql \: // localhost \: 3306/test? user \ = root&password \ = 123456&useUnicode\=true&characterEncoding\=UTF-8

创建 log4j. properties 文件，内容如下。

```
### set log levels ###
log4j. rootLogger = debug, stdout, E, D
log4j. appender. stdout = org. apache. log4j. ConsoleAppender
log4j. appender. stdout. Target = System. out
log4j. appender. stdout. layout = org. apache. log4j. PatternLayout
log4j. appender. stdout. layout. ConversionPattern=%d [%t] %-5p %c - %m%n
log4j. appender. D = org. apache. log4j. DailyRollingFileAppender
log4j. appender. D. File = ${catalina. base}/logs/__TAMS__debug__. log
log4j. appender. D. Append = true
log4j. appender. D. Threshold = ERROE
log4j. appender. D. layout = org. apache. log4j. PatternLayout
log4j. appender. D. layout. ConversionPattern = %-d{yyyy-MM-dd HH\:mm\:ss}
[%t\:%r] - [%p]%m%n
log4j. appender. E = org. apache. log4j. DailyRollingFileAppender
log4j. appender. E. File = ${catalina. base}/logs/__TAMS__error__. log
log4j. appender. E. Append = true
log4j. appender. E. Threshold = ERROE
log4j. appender. E. layout = org. apache. log4j. PatternLayout
log4j. appender. E. layout. ConversionPattern =%-d{yyyy-MM-dd HH\:mm\:ss}
[%t\:%r] - [%p]%m%n
log4j. logger. org. springframework=ERROR
log4j. logger. com. opensymphony=ERROR
log4j. logger. org. apache=ERROR
```

```
log4j.logger.java.sql.ResultSet=INFO
log4j.logger.org.directwebremoting=INFO
```

步骤 4:按 MVC 框架建立不同业务层的包,如图 6-5 所示。

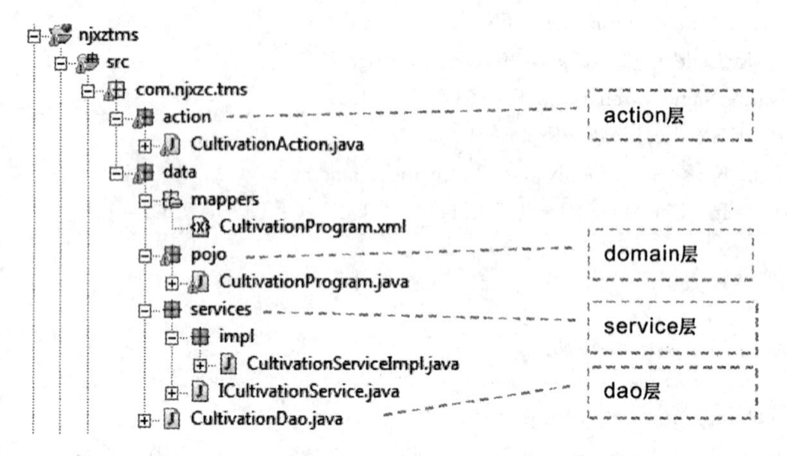

图 6-5 创建业务层

步骤 5:在 com.njxzc.tms.data.pojo 包中先建与表 cultivationprogram 对应的 Java 类 POJO,代码如下。

```
package com.njxzc.tms.data.pojo;
import java.util.Date;
import java.util.List;
public class CultivationProgram {
    private int cultivationProgramId; // 培养方案 ID
    private String cultivationTarget; // 培养目标内容
    private String cultivationSpecification; // 培养规格内容
    private String schoolingPeriod; // 学制内容
    private String degreeName; // 学位内容
    private String graduationRequirement; // 毕业具体条件
    private String creator; // 创建人
    private Date createTime; // 创建时间
    private int schoolYear; // 入学年份
    private String professionCode; // 专业(方向)代码
    private String cultivationName; // 培养方案名称
    public String getCultivationName() {
        return cultivationName;
    }
    public void setCultivationName(String cultivationName) {
        this.cultivationName = cultivationName;
    }
    public int getCultivationProgramId() {
```

```
        return this. cultivationProgramId;
    }
 ⋮
set get 方法省略

}}
```

在com. njxzc. tms. data 包中创建操作 POJO 对象的接口，代码如下。

```
package com. njxzc. tms. data;
import com. njxzc. tms. data. pojo. CultivationProgram;
import java. util. List;
public interface CultivationDao {
    public void saveCultivation(CultivationProgram culProgram); // 保存培养方案
    public void updateCultivation(CultivationProgram culProgram);
    public void delCultivation(int culProgramId);
    public List<CultivationProgram> findAllCulInfos();

}
```

步骤 6：建立 POJO 类与对应表关联配置文件，首先建立两个包，分别是 mybatis. config 和 ibatis. mapper，在 mybatis. config 中建立 mybatis-config. xml 文件，代码如下。

```
<? xml version="1. 0" encoding="UTF - 8" ? >
<! DOCTYPE configuration PUBLIC "-//mybatis. org//DTD Config 3. 0//EN" "http://mybatis.
org/dtd/mybatis - 3 - config. dtd">
<configuration>

    <typeAliases>
            < typeAlias  alias = " CultivationProgram "  type = " com. njxzc. tms. data. pojo.
CultivationProgram" />
    </typeAliases>
    <mappers>
        <mapper resource="com/njxzc/tms/data/mappers/CultivationProgram. xml" />

    </mappers>
</configuration>
```

在 mybatis. mapper 包中创建 CultivationProgram. xml 文件，代码如下。

```
<? xml version="1. 0" encoding="UTF - 8" ? >
<! DOCTYPE mapper PUBLIC "-//mybatis. org//DTD Mapper 3. 0//EN"
      "http://mybatis. org/dtd/mybatis - 3 - mapper. dtd">
<mapper namespace="com. njxzc. tms. data. CultivationDao">
```

```
<! …设置返回集合类型…>
<resultMap type="CultivationProgram" id="RCP">
    <result property="cultivationProgramId"
        column="cultivationProgramId" />
    <result property="cultivationTarget" column="cultivationTarget" />

    <result property="cultivationSpecification"
        column="cultivationSpecification" />
    <result property="schoolingPeriod" column="schoolingPeriod" />
    <result property="degreeName" column="degreeName" />
    <result property="graduationRequirement"
        column="graduationRequirement" />
    <result property="creator" column="creator" />
    <result property="createTime" column="createTime" />
    <result property="schoolYear" column="schoolYear" />
    <result property="professionCode" column="professionCode" />
    <result property="cultivationName" column="cultivationName" />
</resultMap>
<! …具体执行的 SqL 语句…>
<select id="findAllCulInfos" resultMap="RCP">
    SELECT * FROM cultivationprogram
</select>
<insert id="saveCultivation" parameterType="CultivationProgram">
    <![CDATA[
    insert into cultivationProgram
(cultivationTarget,cultivationSpecification,schoolingPeriod,
      degreeName,graduationRequirement,creator,schoolYear,
      professionCode,cultivationName)
      values

(#{cultivationTarget},#{cultivationSpecification},#{schoolingPeriod},

#{degreeName},#{graduationRequirement},#{creator},#{schoolYear},
        #{professionCode},#{cultivationName})
    ]]>
</insert>
<update id="updateCultivation" parameterType="CultivationProgram">
    <![CDATA[
    update cultivationProgram set

cultivationTarget=#{cultivationTarget},cultivationSpecification=#{cultivationSpecification},
        schoolingPeriod=#{schoolingPeriod}, degreeName=#{degreeName},
```

```
    graduationRequirement = #{graduationRequirement}, creator = #{creator}, schoolYear = #
{schoolYear}, professionCode= #{professionCode},
            cultivationName= #{cultivationName},status= #{status}
             where cultivationProgramId= #{cultivationProgramId}
        ]]>
    </update>
    <delete id="delCultivation" parameterType="int">
        <![CDATA[
        delete from cultivationProgram
            where cultivationProgramId= #{id}
        ]]>
    </delete>
</mapper>
        ]]>
    </update>
    <delete id="delCultivation" parameterType="int">
        <![CDATA[
        delete from cultivationProgram
            where cultivationProgramId= #{id}
        ]]>
    </delete>
</mapper>
```

通过以上两个配置文件，把 dao 层与 domain 层及数据库中的表进行了关联，使得 dao 接口中的方法能通过 POJO 对象的操作映射到对数据库中表内容的操作。

步骤 7：在业务逻辑 com. njxzc. tms. data. service 接口包和 com. njxzc. tms. data. service. impl 实现包中分别创建文件，内容是对 CultivationProgram 对象操作的方法接口及实现。其中，接口文件 ICultivationService. java 的代码如下。

```
package com. njxzc. tms. data. services;
import java. util. List;
import com. njxzc. tms. data. pojo. CultivationProgram;
public interface ICultivationService {
        public void saveCultivation(CultivationProgram culProgram); //保存培养方案
    public void updateCultivation(CultivationProgram culProgram); //修改、更新培养方案
    public void delCultivation(int culProgramId); //根据 id 号删除培养方案
    public List<CultivationProgram> findAllCulInfos(); //查询所有培养方案
}
```

接口文件 ICultivationService. javade 的实现 CultivationServiceImpl. java 代码如下。

```
package com. njxzc. tms. data. services. impl;
import java. util. List;
import org. springframework. beans. factory. annotation. Autowired;
import org. springframework. stereotype. Service;
import com. njxzc. tms. data. CultivationDao;
import com. njxzc. tms. data. pojo. CultivationProgram;
import com. njxzc. tms. data. services. ICultivationService;
@Service("cultivationService")//注册一个服务
public class CultivationServiceImpl implements ICultivationService
{
    @Autowired CultivationDao cultivationDao;
    public void delCultivation(int culProgramId)
    {
        cultivationDao. delCultivation(culProgramId);
    }
    public List<CultivationProgram> findAllCulInfos()
    {
        return cultivationDao. findAllCulInfos();
    }
    public void saveCultivation(CultivationProgram culProgram)
    {
        cultivationDao. saveCultivation(culProgram);
    }
    public void updateCultivation(CultivationProgram culProgram)
    {
        cultivationDao. updateCultivation(culProgram);
    }

}
```

在 CurriculumServiceImpl. java 文件中通过“@service”注解的方式把该业务层的方法发布给 Action 层，同时通过“@ Autowired”注解完成从 spring 配置文件中查找满足 culvitationDao 类型的 bean”注解方式去操纵 dao 层，从而实现对底层数据库的操作。接着就需要配置 Sping 来支持该注解方式的运行。

步骤 8：在 src 下创建 applicationContext-common. xml 文件，代码如下。

```
<? xml version="1. 0" encoding="UTF-8"? >
<beans xmlns="http://www. springframework. org/schema/beans"
        xmlns:xsi="http://www. w3. org/2001/XMLSchema-instance"
        xmlns:aop="http://www. springframework. org/schema/aop"
        xmlns:tx="http://www. springframework. org/schema/tx"
        xmlns:jdbc="http://www. springframework. org/schema/jdbc"
        xmlns:context="http://www. springframework. org/schema/context"
```

```
        xsi:schemaLocation="
        http://www.springframework.org/schema/context
http://www.springframework.org/schema/context/spring-context-3.0.xsd
        http://www.springframework.org/schema/beans
http://www.springframework.org/schema/beans/spring-beans-3.0.xsd
        http://www.springframework.org/schema/jdbc
http://www.springframework.org/schema/jdbc/spring-jdbc-3.0.xsd
        http://www.springframework.org/schema/tx
http://www.springframework.org/schema/tx/spring-tx-3.0.xsd
        http://www.springframework.org/schema/aop
http://www.springframework.org/schema/aop/spring-aop-3.0.xsd"
      default-autowire="byName">
   <context:annotation-config />
      <context:property-placeholder
location="classpath:mysql.properties" />
      <context:component-scan base-package="com.njxzc.tms" />
      <bean id="dataSource"
class="org.springframework.jdbc.datasource.DriverManagerDataSource">
            <property name="driverClassName"
value="${jdbc.driverClassName}" />
            <property name="url" value="${jdbc.url}" />
      </bean>
      <bean id="transactionManager"
class="org.springframework.jdbc.datasource.DataSourceTransactionManager">
            <property name="dataSource" ref="dataSource" />
      </bean>
      <bean id="sqlSessionFactory"
class="org.mybatis.spring.SqlSessionFactoryBean">
            <property name="configLocation"
value="classpath:mybatis-config.xml" />
            <property name="dataSource" ref="dataSource" />
      </bean>
      <bean id="cultivationDao"
class="org.mybatis.spring.MapperFactoryBean">
            <property name="sqlSessionFactory" ref="sqlSessionFactory" />
            <property name="mapperInterface"
value="com.njxzc.tms.data.CultivationDao" />
      </bean>
</beans>
```

修改 web.xml 文件，内容如下。

```
<? xml version="1.0" encoding="UTF-8"? >
<web-app version="2.4" xmlns="http://java.sun.com/xml    /ns/j2ee"
    xmlns:xsi="http://www.w3.org/2001/XMLSchema-instance"
    xsi:schemaLocation="http://java.sun.com/xml/ns/j2ee
http://java.sun.com/xml/ns/j2ee/web-app_2_4.xsd">
    <listener>
    <listener-class>org.springframework.web.context.ContextLoaderListener</listener-class>
    </listener>
    <context-param>
        <param-name>contextConfigLocation</param-name>
    <param-value>/WEB-INF/applicationContext-*.xml,classpath*:applicationContext-
*.xml</param-value>
    </context-param>
      <context-param>
         <param-name>log4jConfigLocation</param-name>
         <param-value>/WEB-INF/log4j.properties</param-value>
      </context-param>
      <context-param>
         <param-name>log4jRefreshInterval</param-name>
         <param-value>60000</param-value>
      </context-param>
      <listener>
    <listener-class>org.springframework.web.util.Log4jConfigListener</listener-class>
      </listener>
    <filter>
        <filter-name>struts-cleanup</filter-name>
        <filter-class>
            org.apache.struts2.dispatcher.ActionContextCleanUp
        </filter-class>
    </filter>
    <filter>
        <filter-name>struts2.0</filter-name>
    <filter-class>org.apache.struts2.dispatcher.FilterDispatcher</filter-class>
    </filter>
    <filter-mapping>
        <filter-name>struts-cleanup</filter-name>
        <url-pattern>*.do</url-pattern>
        <url-pattern>*.htm</url-pattern>
        <url-pattern>*.action</url-pattern>
        <url-pattern>*.jsp</url-pattern>
    </filter-mapping>
    <filter-mapping>
        <filter-name>struts2.0</filter-name>
```

```
        <url-pattern>*.do</url-pattern>
        <url-pattern>*.htm</url-pattern>
        <url-pattern>*.action</url-pattern>
        <url-pattern>*.jsp</url-pattern>
    </filter-mapping>
    <servlet>
        <servlet-name>RandImg</servlet-name>
        <servlet-class>com.web.PictureCheckCode</servlet-class>
    </servlet>
    <servlet-mapping>
        <servlet-name>RandImg</servlet-name>
        <url-pattern>/RandImg</url-pattern>
    </servlet-mapping>
    <welcome-file-list>
        <welcome-file>index.jsp</welcome-file>
    </welcome-file-list>
    <login-config>
        <auth-method>BASIC</auth-method>
    </login-config>
</web-app>
```

步骤 9：在 com.njxzc.tms.action 包中创建 CultivationAction.java 文件，代码如下。

```
package com.njxzc.tms.action;
import java.util.List;
import org.springframework.beans.factory.annotation.Autowired;
import com.njxzc.tms.data.pojo.CultivationProgram;
import com.njxzc.tms.data.services.ICultivationService;
public class CultivationAction
{
    @Autowired
    // @Auotowired 和@Resource(name = "cultivationService")
类似 cultivationService 为 CultivationServiceImpl.java 中定义
    ICultivationService culService;
    private CultivationProgram cul;
    private String culId;
    private String Id;
    private List culList;
    private String tips;
    public void setCulService(ICultivationService culService)
    {
        this.culService = culService;
    }
```

```
public ICultivationService getCulService()
{
    return culService;
}
public void setCul(CultivationProgram cul)
{
    this.cul = cul;
}
public CultivationProgram getCul()
{
    return cul;
}
public void setCulId(String culId)
{
    this.culId = culId;
}
public String getCulId()
{
    return culId;
}
public void setId(String id)
{
    Id = id;
}
public String getId()
{
    return Id;
}
public String getTips()
{
    return tips;
}
public void setTips(String tips)
{
    this.tips = tips;
}
public void setCulList(List culList)
{
    this.culList = culList;
}
public List getCulList()
{
    return culList;
```

```
}
public String addCultivation()
{
    String result = "error";
    try
    {
        culService.saveCultivation(cul);
        this.setTips("添加成功");
        result = "success";
    }
    catch (Exception e)
    {
        e.printStackTrace();
        this.setTips("系统出现问题");
    }
    return result;
}
public String viewCulvitation()
{
    String result = "error";
    try
    {
        setCulList(culService.findAllCulInfos());
        result = "success";
    }
    catch (Exception e)
    {
        e.printStackTrace();
        this.setTips("系统出现问题,请稍后访问");
    }
    return result;
}
public String updateCulvitation()
{
    String result = "error";
    try
    {
        culService.updateCultivation(getCul());
        result = "success";
    }
    catch (Exception e)
    {
        e.printStackTrace();
```

```
            this. setTips("更新操作失败");
        }
        return result;
    }
    public String removeCulvitation()
    {
        String result = "error";
        System. out. println("id:" + this. getCulId());
        try
        {
culService. delCultivation(Integer. parseInt(this. getCulId()));
          System. out. println("id:" + this. getCulId());
          result = "success";
        }
        catch (Exception e)
        {
            e. printStackTrace();
            this. setTips("删除操作失败");
        }
        return result;
    }

}
```

在该文件中，通过注解@Autowired 方式调用 service 层的方法，在 strtuts. xml 中配置 action 与 jsp 文件的对应关系完成 jsp 或 action 调用。src 中 struts. xml 文件如下。

```
<? xml version="1. 0" encoding="UTF-8" ? >
<! DOCTYPE struts PUBLIC
    "-//Apache Software Foundation//DTD Struts Configuration 2. 0//EN"
    "http://struts. apache. org/dtds/struts-2. 0. dtd">
<struts>
    <include file="struts-default. xml" />
    <package name="cultivation" extends="struts-default" >
    <action name="addCultivation"
class="com. njxzc. tms. action. CultivationAction"
method="addCultivation">
            <result name="success">/success. jsp</result>
            <result name="error">/error. jsp</result>

    </action>
    <action name="viewCulvitation"
class="com. njxzc. tms. action. CultivationAction"
```

```
method="viewCulvitation">
                <result name="success" >/viewCulvitation.jsp</result>
                <result name="error">/viewCulvitation.jsp</result>
        </action>
        <action name="removeCulvitation"
class="com.njxzc.tms.action.CultivationAction"
method="removeCulvitation">
                <result name="success"
type="redirectAction">/viewCulvitation.action</result>
                <result name="error"
type="redirectAction">/viewCulvitation.action</result>
        </action>
        </package>
</struts>
```

创建index.jsp文件实现主页显示，具体代码如下。

```
<%@ page language="java" pageEncoding="UTF-8"%>
<! DOCTYPE HTML PUBLIC "-//W3C//DTD HTML 4.01 Transitional//EN">
<html>
  <head>
    <title>培养方案管理页面</title>
    <meta http-equiv="pragma" content="no-cache">
    <meta http-equiv="cache-control" content="no-cache">
    <meta http-equiv="expires" content="0">
    <meta http-equiv="keywords" content="keyword1,keyword2,keyword3">
    <meta http-equiv="description" content="This is my page">
  </head>
  <body>
    <h2><a href="addCulvitation.jsp"> 添加培养方案</a>
      <a href="viewCulvitation.action">浏览培养方案</a></h2>
  </body>
</html>
```

创建viewculvitationl.jsp文件，该文件利用Struts 2标签与Aaction交互，把表单信息提交给Action，代码如下。

```
<%@ page language="java" pageEncoding="UTF-8"%>
<%@ taglib prefix="s" uri="/struts-tags"%>
<! DOCTYPE HTML PUBLIC "-//W3C//DTD HTML 4.01 Transitional//EN">
<html>
    <head>
        <title>浏览培养方案</title>
    </head>
```

```
    <body>
        <table width="85%" border="1" align="center" style="width:80%;">
            <tr>
        <th colspan="9" align="center" valign="middle">培养方案列表</th>
            </tr>
            <tr>
<td colspan="9" align="left" valign="middle"><a
href="<%=request.getContextPath()%>/addCulvitation.jsp"><em>添加培养方案</em>
</a>
</td>
</tr>
<tr>
<td width="30%" align="center" valign="middle"><h4>方案名称</h4>
</td>
<td width="7%" align="center" valign="middle"><h4>学制</h4></td>
<td width="14%" align="center" valign="middle"><h4>学位名称</h4>
</td>
<td width="13%" align="center" valign="middle"><h4>创建人</h4>
</td>
<td width="14%" align="center" valign="middle"><h4>入学年份</h4>
</td>
<td width="22%" align="center" valign="middle"><h4>操作</h4></td>
</tr>
<s:iterator value="culList">
    <tr>
    <td height="23" align="center" valign="middle">
    <s:property value="cultivationName"/>
    </td>
    <td align="center" valign="middle">
    <s:property value="schoolingPeriod"/>
    </td>
    <td align="center" valign="middle">
    <s:property value="degreeName"/>
    </td>
    <td align="center" valign="middle">
    <s:property value="creator"/>
    </td>
     <td align="center" valign="middle">
     <s:property value="schoolYear"/>
      </td>
       <td align="center" valign="middle">
       <a
```

```
href="<%=request.getContextPath()%>/findById.action?culId=${cultivationProgramId}">
修改</a>
        <a
href = "<% = request.getContextPath()%>/removeCulvitation.action?culId =
${cultivationProgramId}">删除</a>
                    </td>
                      </tr>
                </s:iterator>
                <s:property value="tips"/>
            </table>
        </body>
</html>
```

在这个JSP文件中利用Struts 2的特性，使得JSP表单能够直接把对象传递到Action，体现了Struts 2的View层与Action通信的便利性。

最后，项目的层次结构如图6-6所示。

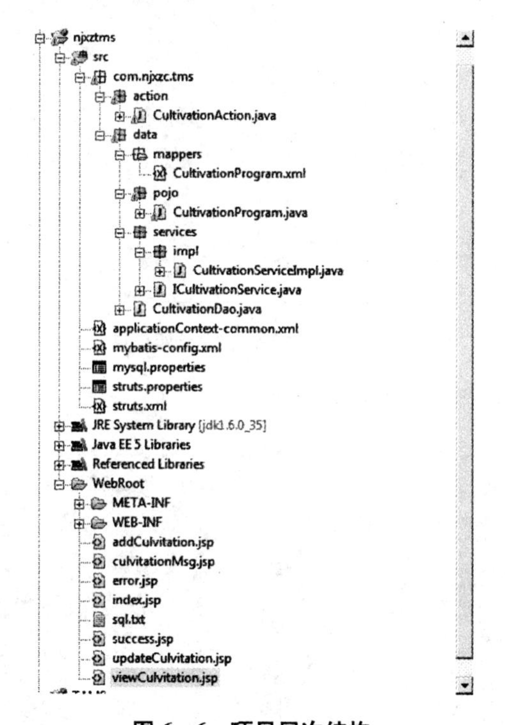

图6-6　项目层次结构

实现效果：在浏览器中执行"http://localhost:8080/njxztms/"，界面如图6-7所示。

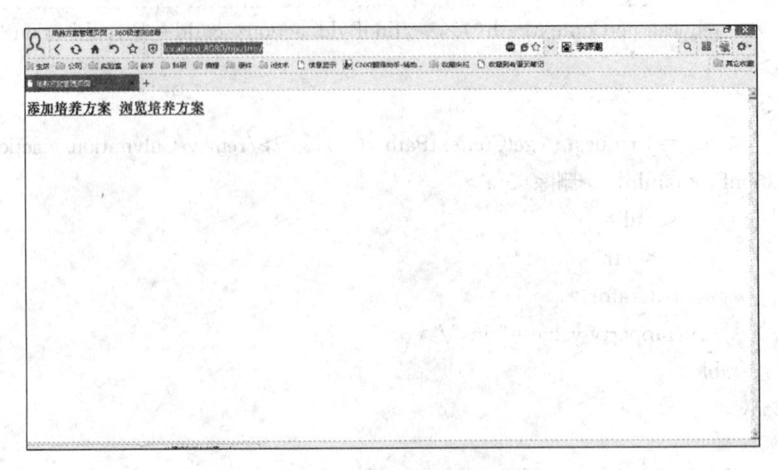

图 6－7　培养方案管理页面

选择创建培养方案,如图 6－8 所示。

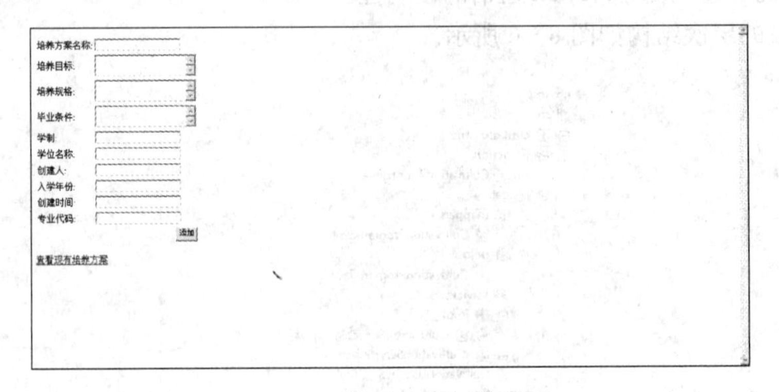

图 6－8　培养方案添加页面

查看所有培养方案,如图 6－9 所示。

培养方案列表

添加培养方案

方案名称	学制	学位名称	创建人	入学年份	操作
2011软件工程(金融)	4	工学学士	王小正	2011	修改 删除
2011软件工程(服务外包)	4	工学学士	潘雷	2011	修改 删除
2011智能终端应用	4	工学学士	杨蔚鸣	2011	修改 删除
2010嵌入式系统	4	工学学士	曲波	2010	修改 删除
2010软件工程本科专业人才培养方案	4	工学学士	潘雷	2010	修改 删除
2011软件工程（软件测试）	4	工学学士	李雯睿	2011	修改 删除
2010行知数据库技术	4	理学学士	杨种学	2010	修改 删除
2009嵌入式系统	4	工学学士	曲波	2009	修改 删除
2009师范	4	理学学士	杨种学	2009	修改 删除
2009软件工程	4	理学学士	杨种学	2009	修改 删除
计算机科学与技术本科专业人才培养方案	4	理学学士	杨种学	2009	修改 删除
2010计算机科学与技术（网络方向）	4	工学学士	杨种学	2010	修改 删除
2010中韩1.2（软件工程）	4	理学学士	杨种学	2010	修改 删除
2011计算机科学与技术(网络工程)	4	工学学士	王江平	2011	修改 删除
2011数学与应用数学（保险精算）	4	工学学士	任艳丽	2011	修改 删除
2013智能终端培养方案	4	工学学士	4徐家喜	2013	修改 删除

图 6－9　培养方案浏览页面

项目最终整体实现效果:

整个项目功能较多,其他功能的实现可以安排学生分组实现。用户登录正确后,进入系统,如图 6－10 所示。

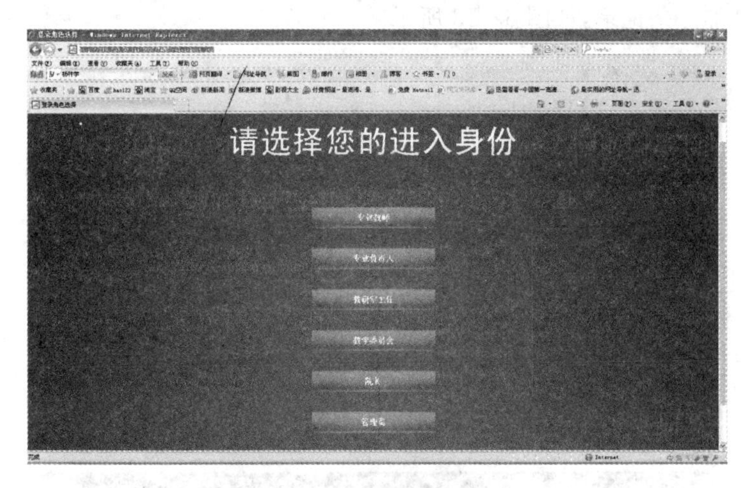

图 6-10　登录后模块选择页面

接着按不同角色进入不同页面，如角色为专业负责人，则进入如图 6-11 所示界面。

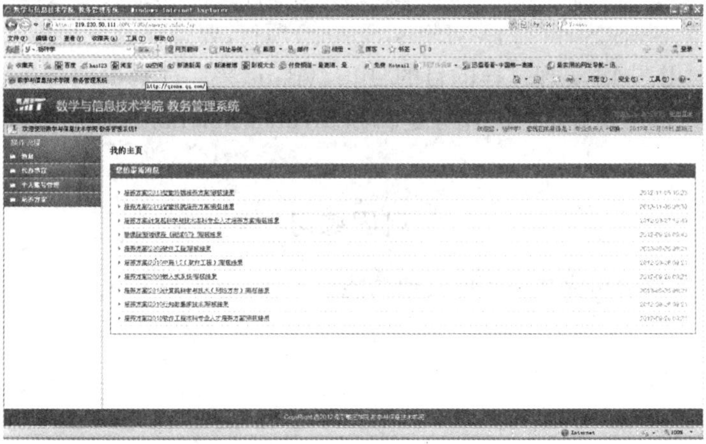

MIT 数学与信息技术学院 教务管理系统

我的主页

图 6-11　专业负责人页面

选择创建培养方案，如图 6-12 所示。

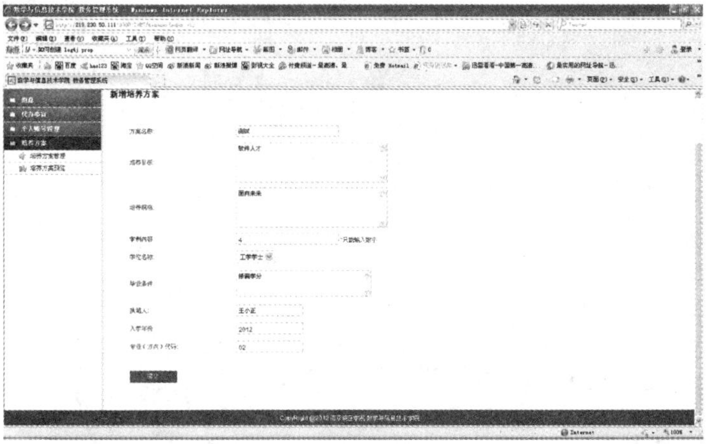

新增培养方案

软件人才

4

工学学士

王小正

2012

02

图 6-12　创建培养方案页面

提交后则增加一条记录，如图 6 - 13 所示。

图 6 - 13 培养方案增加记录页面

从整个操作流程大家可以看出，Java EE 的体系结构有三层：表示层、业务逻辑层和数据持久层。开发一个 SMS(Struts 2，MyBatis，Spring)项目，要遵循这三层模式。根据前面所学的知识，可以分别用 SMS 实现这样的目的：用 MyBatis 来完成数据的持久层应用，用 Spring 的 Bean 来管理组件（主要是 DAO、业务逻辑和 Struts 的 Action），而用 Struts 来完成页面的控制跳转。

巩固练习

1. 根据项目需求分组完成项目的其他功能。

第7章 Spring 3 MVC和Hibernate整合应用
——教育资源平台开发实例

学习目标

1. 了解并掌握Spring 3 MVC和Hibernate两种框架整合应用的配置。
2. 掌握项目开发流程。
3. 巩固所学框架知识。

该项目是由某校师生与企业联合开发出的一个教育资源系统，通过该项目的锻炼，使师生对项目开发过程、项目管理和最新开发框架有了充分了解。

§7.1 项目简介

1. 项目功能架构(如图7-1所示)

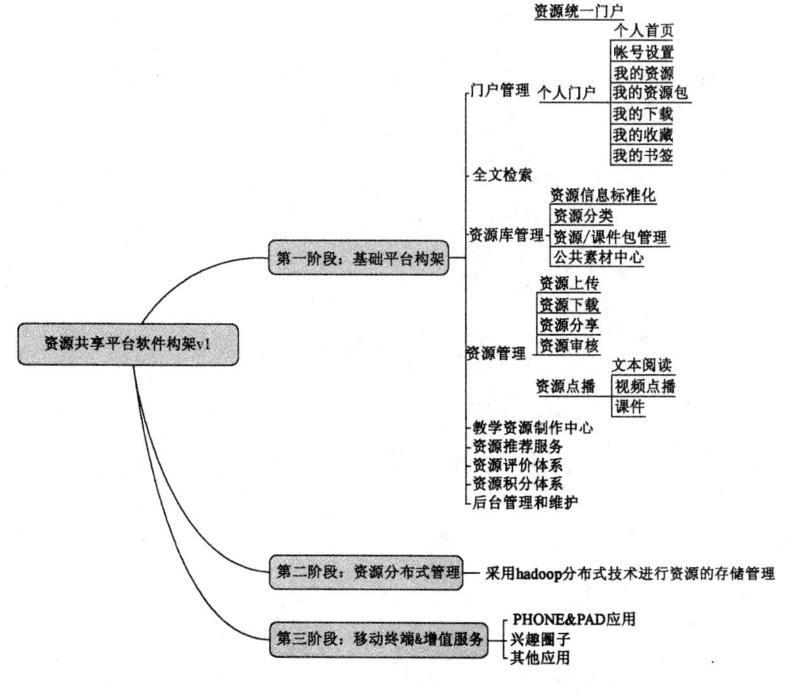

图7-1 平台功能架构

§7.2　技术架构

(1) 平台开发环境

① Spring 3.0 和 Hibernate 3.0 作为底层技术构架；

② Tomcat 7.0；

③ Oracle 10g；

④ JDK7.X；

⑤ Eclipse Java EE IDE for Web Developers(需要安排 maven 和 subeclipse 插件，可以直接打开 Eclipse，Help - Marketplace 中下载)。

(2) 项目编译环境设置

先建一个 web Project，上传到 SVN 服务器。项目组所有成员在项目开发过程中通过 SVN 来进行项目版本控制。

① 平台初始包下载

"https://219.230.55.224/svn/RSP/trunk";

用户名为姓名全拼，密码姓名首字母；

采用 Eclipse SVN 下载初始包。

② 使用 Eclipse 配置项目(使用 Mven 管理)

删除在 SVN 检出的 Eclipse 应用，注意请不要勾选删除本地硬盘代码!，如图 7-2 所示。

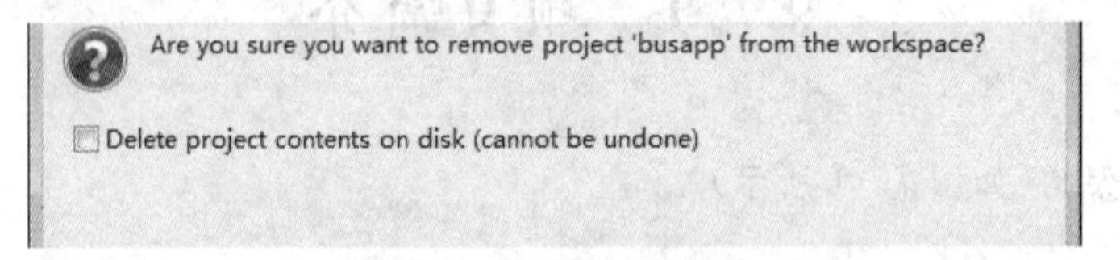

图 7-2　Eclipse 配置

在 Eclipse 中菜单 File - import，选择 Maven - Existing Maven projects，导入之前 SVN 检出的项目，如图 7-3 所示。

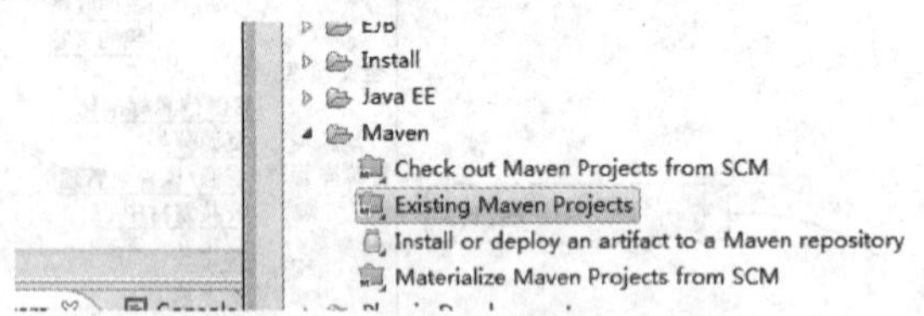

图 7-3　Eclipse 添加 Maven 管理功能

导入成功后项目结构如图 7-4 所示。

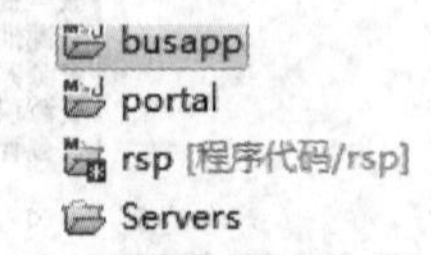

图 7-4　导入成功后项目结构

在 Eclipse 中配置 servers 为 Tomcat 7，如图 7-5 所示。点击 debug 或者 start 运行 Tomcat，如果运行成功则系统搭建和初始化完成。

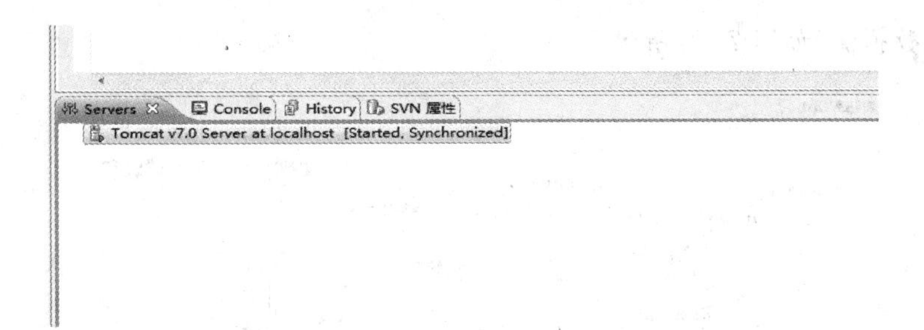

图7-5 配置Tomcat

3. 页面展示

(1) 平台运行后前台首页如图7-6所示。

图7-6 平台运行首页

(2) 管理员后台登陆后页面如图7-7所示。

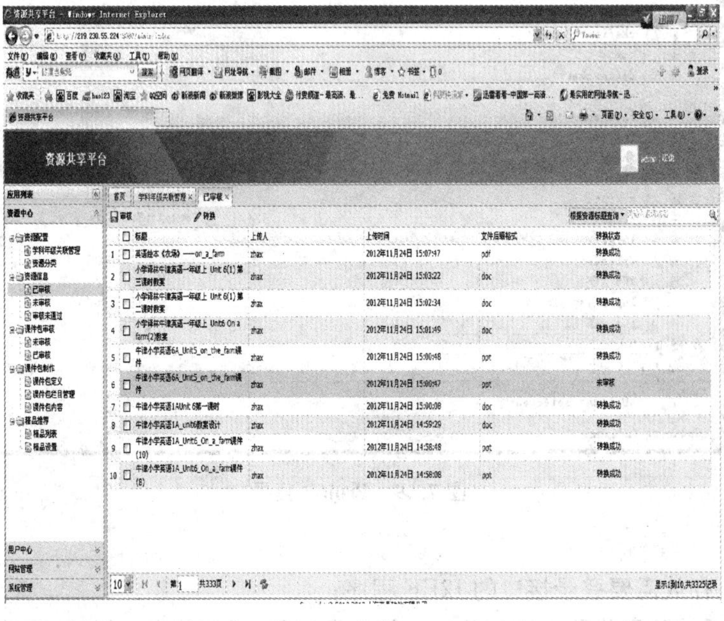

图7-7 后台管理页面

4. 系统数据流(如图 7-8 所示)

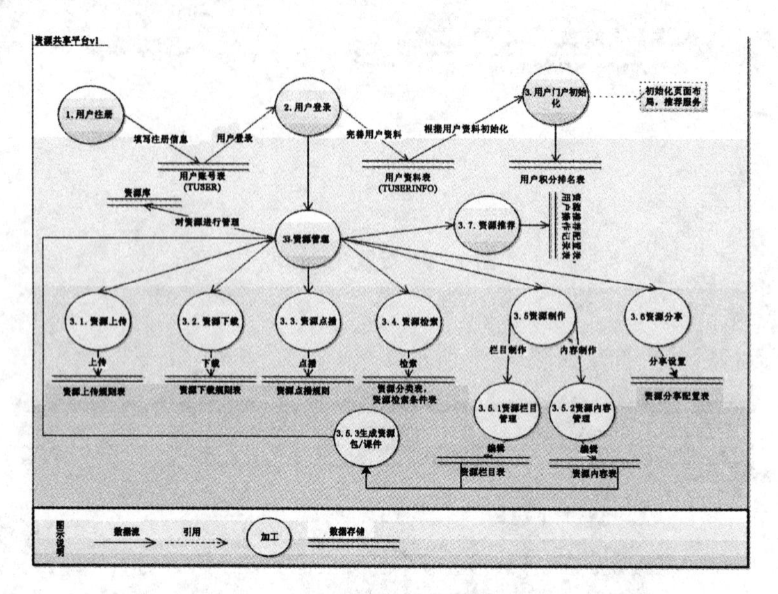

图 7-8　系统数据流

5. Project 目录(如图 7-9 所示)

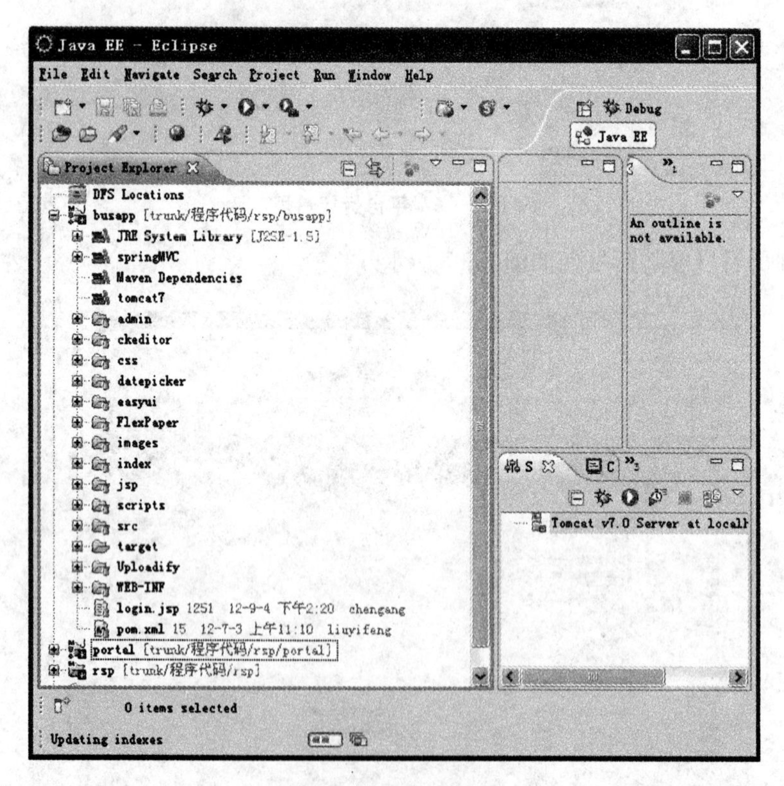

图 7-9　Project 目录

目录说明：

① busapp：存放需要登录接口的 J2EE 程序；

② admin：系统配置优化和监控程序。如数据字段，系统信息，模板管理等；

③ css:系统通用的 css 文件;

④ datepicker:日期控件;

⑤ easyui:页面 UI 库;

⑥ images:界面图片库;

⑦ jsp:J2EE 程序;

⑧ src:J2ee 标准目录,存放配置文件和源码;

⑨ WEB-INF:J2EE 标准程序目录,存放编译后的配置文件和代码,及相关 jar 包;

⑩ portal:存放外部网站和门户。

6. 平台关键 API

(1) Package

① com. rsp. core,存放平台核心包;

② com. rsp. util,平台工具包。

(2) Class

① com. rsp. core. Page,分页处理;

② com. rsp. core. BaseController,action 基类;

③ com. rsp. core. BaseDao,数据处理基类;

④ com. rsp. core. BaseDomain,实体对象基类。

(3) Method

① com. rsp. core. BaseDao. load(Serializable id);

② com. rsp. core. BaseDao. get(Serializable id);

③ com. rsp. core. BaseDao. loadAll;

④ com. rsp. core. BaseDao. save(T entity);

⑤ com. rsp. core. BaseDao. delete(T entity);

⑥ com. rsp. core. BaseDao. deleteAll(T entity);

⑦ com. rsp. core. BaseDao. update(T entity);

⑧ com. rsp. core. BaseDao. find(String hql);

⑨ pagedQuery(String hql, int pageNo, int pageSize, Object... values)。

7. 数据字典(以资源分类表 7-1 为例)

表 7-1　资源分类表(RS_CATEGORY)

字段名称	字段含义	数据类型	宽度	NULL	注
ID	表的主键	NUMBER	12	N	记录的唯一标识
CODE	分类代码	VARCHAR2	100	Y	
CETEGORYNAME	分类名称	VARCHAR2	100	Y	
PARENTID	父节点 ID	NUMBER	12	Y	
RANK	排名	NUMBER	12	Y	
STATUS	开启状态	NUMBER	12	Y	

8. 开发过程(如图 7-10 所示)

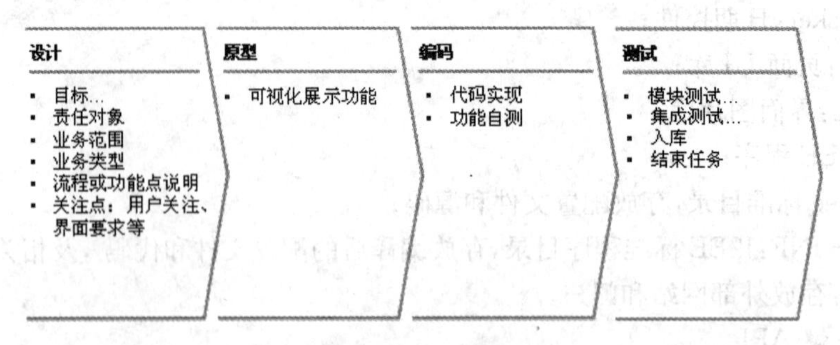

图 7-10　开发过程

9. 开发任务安排(如图 7-11 所示)

开发人员	任务列表	时间要求	截至是时间
伊昭荣	角色分类，角色管理	1D	2012/7/5下班
施俊霞	用户管理，系统参数	1D	2012/7/5下班
岳国宾	权限管理、岗位成员表	1D	2012/7/5下班
张辉	数据字典，学校管理	1D	2012/7/5下班
张颖	组织机构管理	1D	2012/7/5下班

图 7-11

§7.3　两个项目实现技术比较

1. 相同点

项目整体 MVC 框架的层次结构与前一个项目大体一致，逻辑层次也分为 Domain、Dao、Service、Controller 及 JSP，其中 Controller 相当与教务管理系统中的 Action，结构如图 7-12 所示。

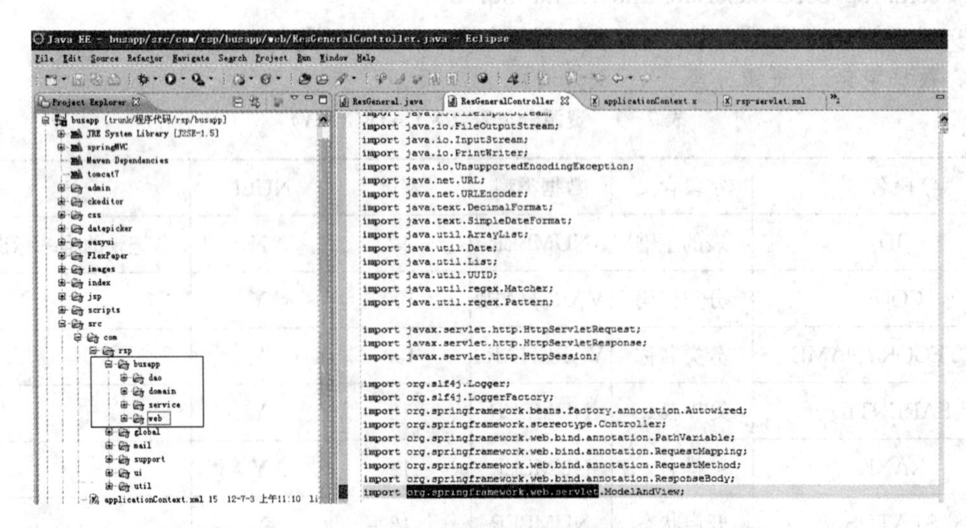

图 7-12　项目层次结构

2. 不同点

(1) 在教务管理系统中，Spring 只用到自身的依赖注入特性，把 Struts 2 的 Action 与底层的 Dao 层连接起来，View 层与 Action 通信是由 Struts 2 完成的。

在教育资源系统中，Spring 3 不仅使用了依赖注入的特性，还充分利用了 Spring 3 MVC 带来的新特性，因此配置时多增加了一个名为 org. springframework. web. servlet - 3. 1. 1. RELEASE. jar 的包文件。

(2) 在数据持久层方面也使用了 JPA Annotations 和 Hibernate Annotations 特性，使用 Hibernate Annotations，无需使用 HBM 映射档案，可直接在 POJO 上使用 Annotation 设定对应关系。

相比较而言，该项目比第 6 章例举的项目配置文件更简洁点，该项目的数据字典见附录。

§7.4　项目创建流程(以用户注册为例)

1. 项目目录结构(如图 7 - 13 所示)

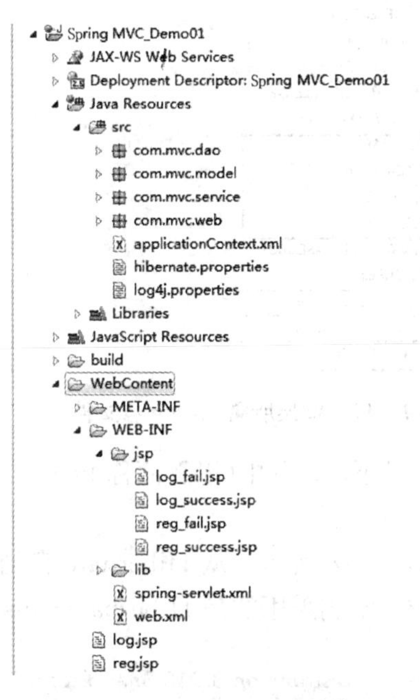

图 7 - 13　项目结构图

2. 项目创建步骤

第一步：添加框架必须的 jar 包。

(1) Spring 中的 jar 包

首先从 Spring 官网“http: // www. springsource. org/download/community”中下载最新版本的 Spring 版本，本例采用的是 Spring - framework - 3. 1. 1. RELEASE。初学者最好采用和本例一样的版本。下载后如图 7 - 14 所示。

Spring-framework-3.1.1.RELEASE-with-docs.zip

图 7 - 14　spring 压缩包

解压得到文件夹，如图 7－15 所示。

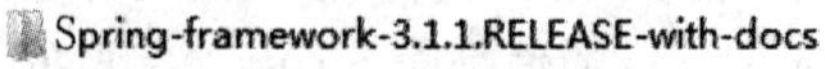

Spring-framework-3.1.1.RELEASE-with-docs

图 7－15 spring 解压缩文件夹

打开文件夹，如图 7－16 所示。

dist
docs
projects
src
changelog.txt
license.txt
notice.txt
readme.txt

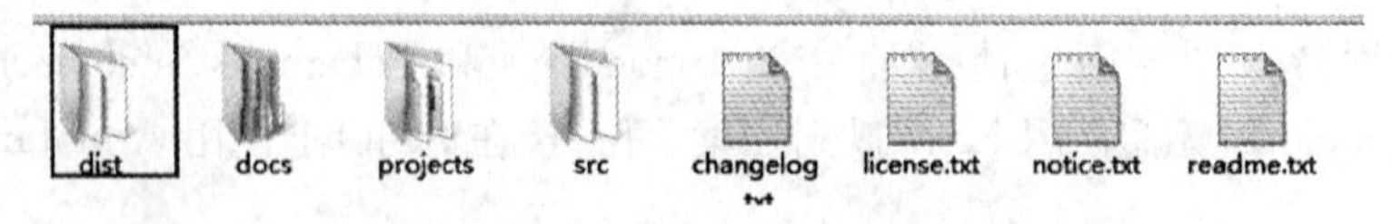

图 7－16 Spring 文件夹

选择如下 jar 包，如图 7－17 所示。

org.springframework.aop-3.1.1.RELEASE.jar
org.springframework.asm-3.1.1.RELEASE.jar
org.springframework.aspects-3.1.1.RELEASE.jar
org.springframework.beans-3.1.1.RELEASE.jar
org.springframework.context.support-3.1.1.RELEASE.jar
org.springframework.context-3.1.1.RELEASE.jar
org.springframework.core-3.1.1.RELEASE.jar
org.springframework.expression-3.1.1.RELEASE.jar
org.springframework.instrument.tomcat-3.1.1.RELEASE.jar
org.springframework.instrument-3.1.1.RELEASE.jar
org.springframework.jdbc-3.1.1.RELEASE.jar
org.springframework.jms-3.1.1.RELEASE.jar
org.springframework.orm-3.1.1.RELEASE.jar
org.springframework.oxm-3.1.1.RELEASE.jar
org.springframework.spring-library-3.1.1.RELEASE.libd
org.springframework.test-3.1.1.RELEASE.jar
org.springframework.transaction-3.1.1.RELEASE.jar
org.springframework.web.portlet-3.1.1.RELEASE.jar
org.springframework.web.servlet-3.1.1.RELEASE.jar
org.springframework.web.struts-3.1.1.RELEASE.jar
org.springframework.web-3.1.1.RELEASE.jar

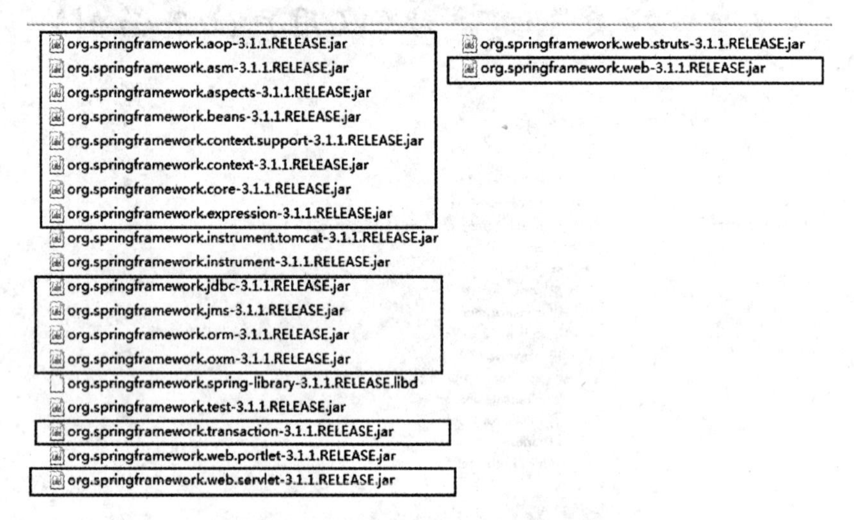

图 7－17 框架所需 Spring 包文件

将以上 jar 包复制到/web－INF 下的/lib 中(以下简称 lib)。

(2) Hibernate 中的 jar 包

准备 Hibernate 的 jar 包，如图 7－18 所示。从 Hibernate 官网"http://www.hibernate.org/downloads"下载 hibernate 版本，本例采用的是 Hibernate－distribution－3.6.10.Final。

Hibernate-distribution-3.6.10.Final-dist.zip

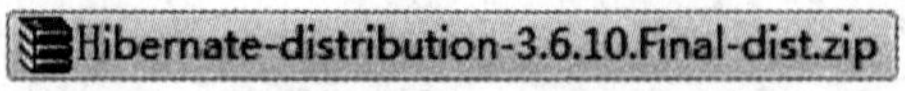

图 7－18 Hibernate 压缩包文件

解压后如图 7－19 所示。

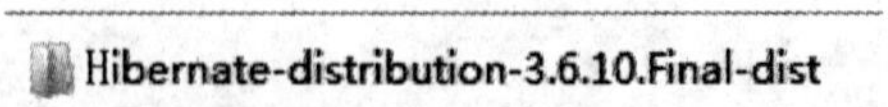

Hibernate-distribution-3.6.10.Final-dist

图 7－19 Hibernate 解压缩文件夹

打开 Hibernate 文件夹，先选择如图 7－20 中的 jar 包。

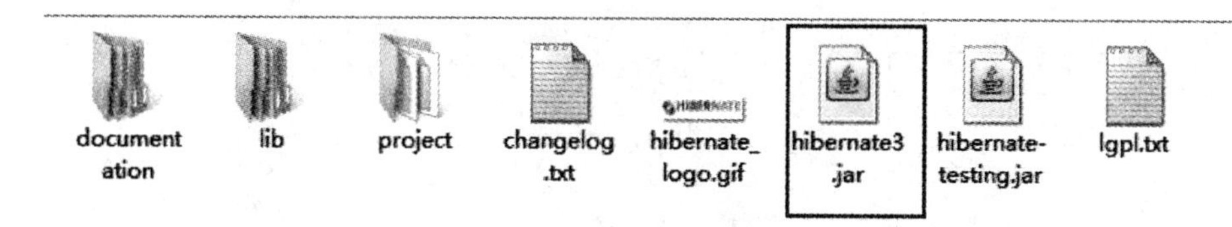

图 7-20　Hibernate 文件夹

再打开 Hibernate 文件夹中的 lib 文件夹，如图 7-21 所示。

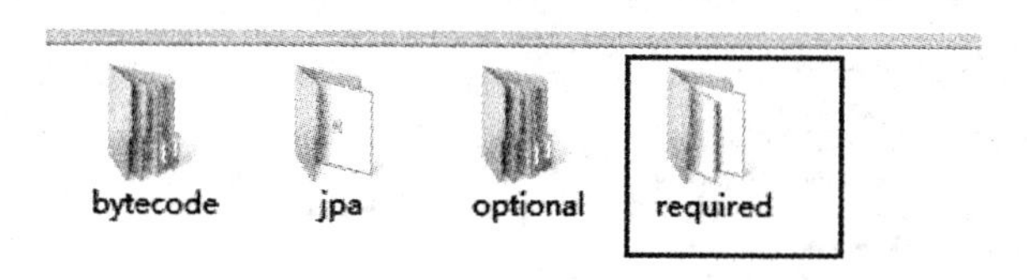

图 7-21　Hibernate 中的 lib 文件夹

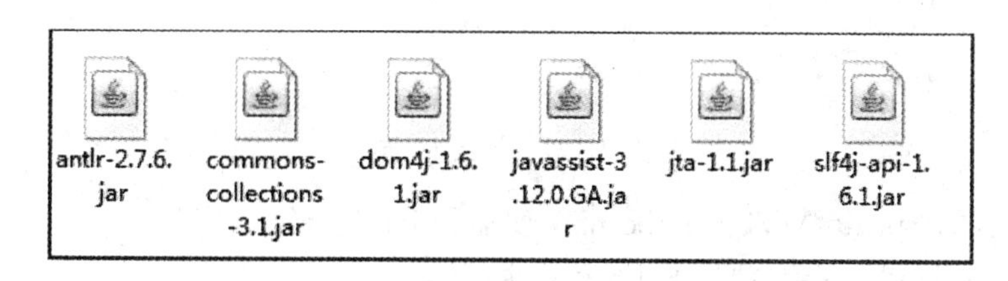

图 7-22　框架所需 Spring 包文件

将此文件夹的 jar 包全部复制到 web 项目的 lib 目录中并将 jpa 中的 jar 也复制到 lib 中。将 bytecode 中的 cglib 复制到 lib 中。分别如图 7-23、7-24、7-25、7-26 所示。

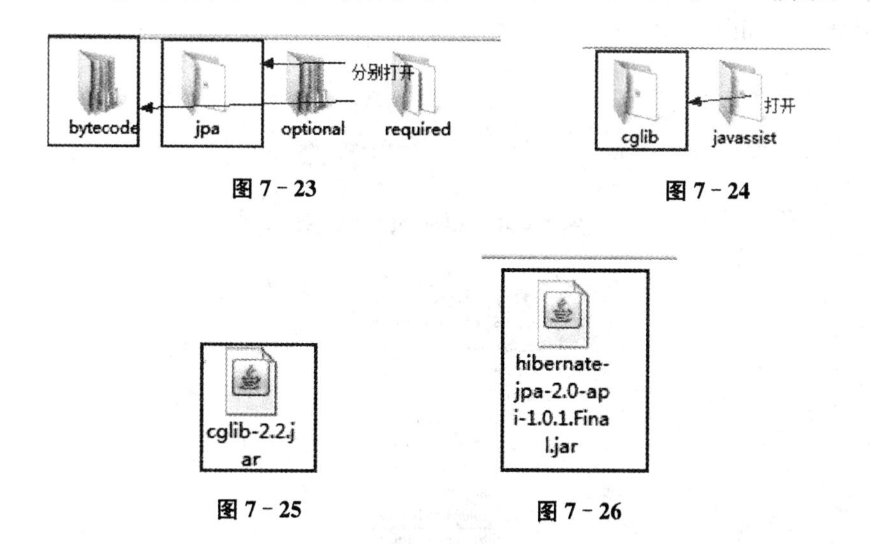

图 7-23

图 7-24

图 7-25

图 7-26

以上是 Hibernate 和 Spring 中所要加入的包，除这些之外还要加入如图 7-27 所示的其余包。

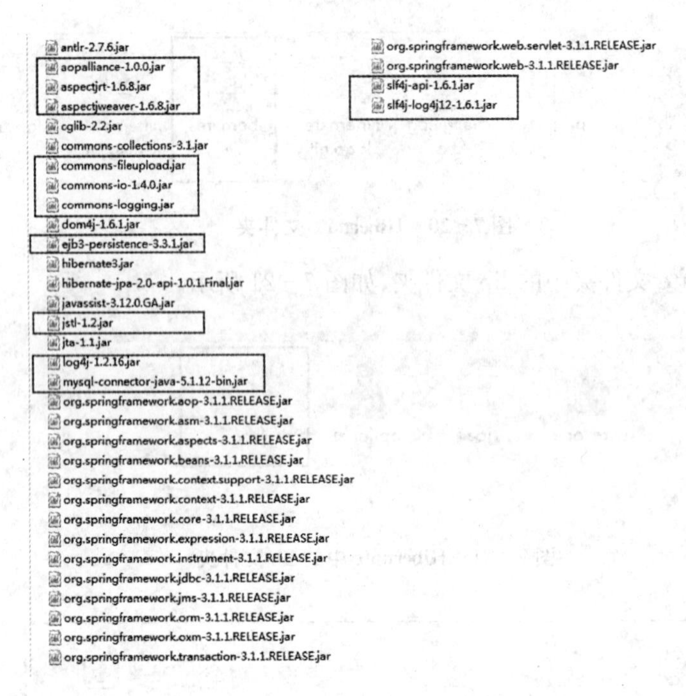

图 7-27　其余 jar 包

以上是所有的 Spring MVC＋Hibernate 所需的 jar 包。

将所有 jar 包复制到 web 项目的 lib 目录中。

第二步：配置文件。

(1) 从 Hibernate 中复制并修改文件

首先从 Hibernate 中的 project 文件夹中打开 etc 文件夹，复制其中如下图所示的两个文件到 classpath 下，如图 7-28、7-29 所示。

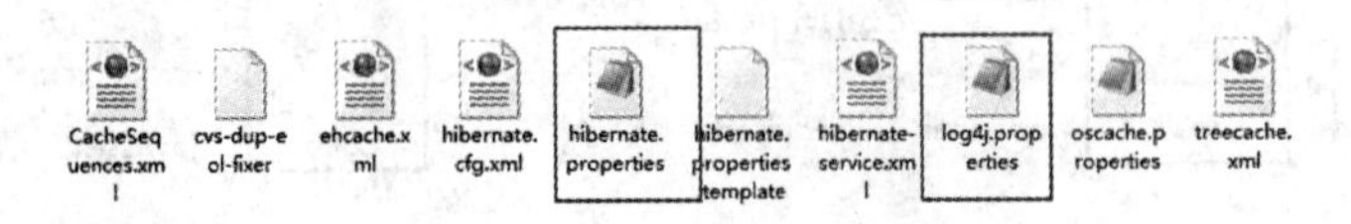

图 7-28　Hibernat 配置文件

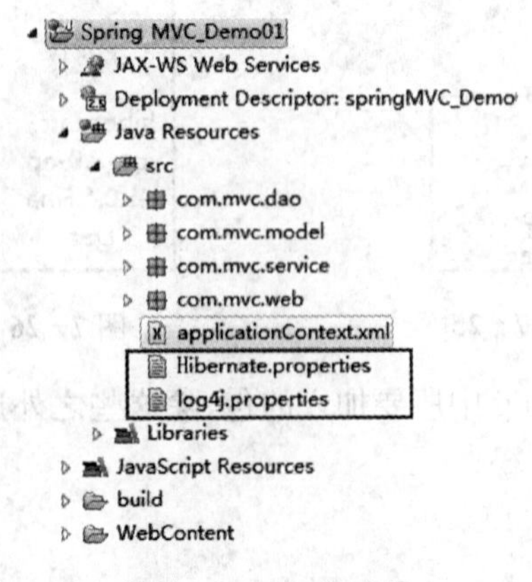

图 7-29　jar 包路径

① 修改 Hibernate. properties 文件

如下：

```
##HypersonicSQL
dataSource.password=root
dataSource.username=root
dataSource.databaseName=test
dataSource.driverClassName=com.mysql.jdbc.Driver
dataSource.dialect=org.hibernate.dialect.MySQL5Dialect
dataSource.serverName=localhost:3306
dataSource.url=jdbc:mysql://localhost:3306/test
dataSource.properties=user=${dataSource.username};
databaseName=${dataSource.databaseName};
serverName=${dataSource.serverName};password=${dataSource.password}
dataSource.hbm2ddl.auto=update
#hibernate.connection.url jdbc:hsqldb:hsql://localhost
#hibernate.connection.url jdbc:hsqldb:test
```

② 修改 log4j. properties 文件

如下：

```
log4j.rootLogger=warn, stdout
#log4j.logger.org.hibernate=info
#log4j.logger.org.hibernate=debug
```

以上两个文件未显示出来的代码表示不改。

(2) 编写 spring 配置文件

① 新建 applicationContext. xml 文件

在 classpath 下新建 applicationContext. xml 文件内容如下：

```
<?xml version="1.0" encoding="UTF-8"?>
<beans xmlns="http://www.springframework.org/schema/beans"
      xmlns:aop="http://www.springframework.org/schema/aop"
xmlns:context="http://www.springframework.org/schema/context"
      xmlns:p="http://www.springframework.org/schema/p"
xmlns:tx="http://www.springframework.org/schema/tx"
      xmlns:xsi="http://www.w3.org/2001/XMLSchema-instance"
      xsi:schemaLocation="
            http://www.springframework.org/schema/beans
http://www.springframework.org/schema/beans/spring-beans-3.0.xsd
            http://www.springframework.org/schema/context
http://www.springframework.org/schema/context/spring-context-3.0.xsd
            http://www.springframework.org/schema/aop
http://www.springframework.org/schema/aop/spring-aop-3.0.xsd
```

```
            http://www.springframework.org/schema/tx
http://www.springframework.org/schema/tx/spring-tx-3.0.xsd">

    <context:component-scan base-package="com.mvc" />
    <!-- 支持 aop 注解 -->
    <aop:aspectj-autoproxy />
    <context:property-placeholder location="classpath:/hibernate.properties" />

    <bean id="dataSource"
          class="org.springframework.jdbc.datasource.DriverManagerDataSource">
          <property name="driverClassName" value="${dataSource.driverClassName}" />
          <property name="url" value="${dataSource.url}" />
          <property name="username" value="${dataSource.username}" />
          <property name="password" value="${dataSource.password}" />
    </bean>
    <bean id="sessionFactory"
    class="org.springframework.orm.hibernate3.annotation.AnnotationSessionFactoryBean">
        <property name="dataSource" ref="dataSource" />
        <property name="HibernateProperties">
            <props>
                <prop key="Hibernate.dialect">${dataSource.dialect}</prop>
                <prop key="Hibernate.hbm2ddl.auto">${dataSource.hbm2ddl.auto}</
prop>
                <prop key="Hibernate.hbm2ddl.auto">update</prop>
            </props>
        </property>
        <property name="packagesToScan">
            <list>
                <value>com.mvc.model</value><!-- 扫描实体类,也就是平时所说的
model -->
            </list>
        </property>
    </bean>
    <bean id="hibernateTemplate" class="org.springframework.orm.hibernate3.
HibernateTemplate">
        <property name="sessionFactory" ref="sessionFactory"></property>
    </bean>

<!-- 配置事务管理 -->
    <bean id="txManager"
        class="org.springframework.orm.hibernate3.HibernateTransactionManager">
        <property name="sessionFactory" ref="sessionFactory"></property>
    </bean>
```

```
    <tx:annotation-driven transaction-manager="txManager" />
    <aop:config>
        <aop:pointcut expression="execution(public * com.mvc.service.impl.*.*(..))"
            id="businessService" />
        <aop:advisor advice-ref="txAdvice" pointcut-ref="businessService" />
    </aop:config>
    <tx:advice id="txAdvice" transaction-manager="txManager">
        <tx:attributes>
                <tx: method name = " find * " read-only = " true" propagation = " NOT_
SUPPORTED" />
              <! -- get 开头的方法不需要在事务中运行 。有些情况是没有必要使用事务的，
比如获取数据。开启事务本身对性能是有一定的影响 -->
              <tx:method name=" * " />      <! -- 其他方法在事务中运行 -->
        </tx:attributes>
    </tx:advice>
</beans>
```

② 新建 spring-servlet. xml 文件

在 web - INF 下新建 spring-servlet. xml 文件，内容如下：

```
<? xml version="1.0" encoding="UTF-8"? >
<beans xmlns="http://www.springframework.org/schema/beans"
xmlns:xsi="http://www.w3.org/2001/XMLSchema-instance"
    xmlns:p="http://www.springframework.org/schema/p"
xmlns:mvc="http://www.springframework.org/schema/mvc"
    xmlns:context="http://www.springframework.org/schema/context"
xmlns:util="http://www.springframework.org/schema/util"
xsi:schemaLocation="http://www.springframework.org/schema/beans
http://www.springframework.org/schema/beans/spring-beans-3.0.xsd
http://www.springframework.org/schema/context
http://www.springframework.org/schema/context/spring-context-3.0.xsd
http://www.springframework.org/schema/mvc
http://www.springframework.org/schema/mvc/spring-mvc-3.0.xsd
http://www.springframework.org/schema/util
http://www.springframework.org/schema/util/spring-util-3.0.xsd">
    <! -- 对 web 包中的所有类进行扫描，以完成 Bean 创建和自动依赖注入的功能 -->
    <context:component-scan base-package="com.mvc.web"/>
    <mvc:annotation-driven />
    <! -- 支持 spring3.0 新的 mvc 注解 -->
    <! -- 启动 Spring MVC 的注解功能，完成请求和注解 POJO 的映射 -->
        < bean class = " org.springframework.web.servlet.mvc.annotation.
AnnotationMethodHandlerAdapter"/>
    <! --对模型视图名称的解析，即在模型视图名称添加前后缀 -->
```

```
    <bean class="org. springframework. web. servlet. view. InternalResourceViewResolver"
          p:prefix="/WEB-INF/jsp/" p:suffix=". jsp">
          <! -- 如果使用 Jstl 的话,配置下面的属性 -->
          <property name="viewClass" value="org. springframework. web. servlet. view. JstlView"
/>
    </bean>
</beans>
```

③ 修改 web. xml 文件

文件内容如下：

```
<? xml version="1. 0" encoding="UTF-8"? >
<web-app version="2. 5"
    xmlns="http://java. sun. com/xml/ns/javaee"
    xmlns:xsi="http://www. w3. org/2001/XMLSchema-instance"
    xsi:schemaLocation="http://java. sun. com/xml/ns/javaee
    http://java. sun. com/xml/ns/javaee/web-app_2_5. xsd">
    <servlet>
          <servlet-name>Spring mvc</servlet-name>
          <servlet-class>   org. springframework. web. servlet. DispatcherServlet
          </servlet-class>
          <init-param>
        <param-name>contextConfigLocation</param-name>
<param-value> classpath: applicationContext *. xml,/WEB-INF/spring-servlet. xml</param-
value>
          </init-param>
          <load-on-startup>1</load-on-startup>
    </servlet>
    <servlet-mapping>
          <servlet-name>springmvc</servlet-name>
          <url-pattern> *. do</url-pattern>
    </servlet-mapping>
    <welcome-file-list>
          <welcome-file>index. *</welcome-file>
          <welcome-file>reg. jsp</welcome-file>
    </welcome-file-list>
</web-app>
```

第三步：业务逻辑

(1) Model 层

新建实体类 User. java，内容如下：

```
package com.mvc.model;

import javax.persistence.Entity;
import javax.persistence.GeneratedValue;
import javax.persistence.Id;

@Entity
public class User {
    private int id;
    private String name;
    private String password;

    @Id
    @GeneratedValue
    public int getId() {
        return id;
    }

    public void setId(int id) {
        this.id = id;
    }

    public String getName() {
        return name;
    }

    public void setName(String name) {
        this.name = name;
    }

    public String getPassword() {
        return password;
    }

    public void setPassword(String password) {
        this.password = password;
    }
}
```

(2) Dao 层

新建 UserDao.java，内容如下：

```
package com. mvc. dao;
import java. util. List;
import javax. annotation. Resource;
import org. springframework. orm. hibernate3. HibernateTemplate;
import org. springframework. stereotype. Repository;
import com. mvc. model. User;
@Repository("userDao")
public class UserDao {
    @Resource
    private HibernateTemplate HibernateTemplate;
    public void add(User u) {
        HibernateTemplate. save(u);
    }

    public boolean findUser(User u) {
        List<? > list = HibernateTemplate. find("from User u where u. name=?",
                u. getName());
        if (list. size() > 0) {
            return true;
        }
        return false;
    }

    public boolean userLogin(User u) {
        List<? > list = HibernateTemplate. find(
          "from User u where u. name=? and u. password=?", u. getName(),
          u. getPassword());
        if (list. size() > 0) {
            return true;
        }
        return false;
    }
}
```

(3) Service 层

新建 UserService. java，内容如下：

```
package com.mvc.service;
import javax.annotation.Resource;
import org.springframework.stereotype.Service;
import com.mvc.dao.UserDao;
import com.mvc.model.User;
@Service("userService")
public class UserService {
    @Resource
    private UserDao userDao;
    public void add(User user) {
        userDao.add(user);
    }
    public boolean exist(User user) {
        return userDao.findUser(user);
    }
    public boolean login(User user) {
        return userDao.userLogin(user);
    }
}
```

(4) Controller 层(Action 层)

新建 UserController.java，内容如下：

```
package com.mvc.web;
import javax.annotation.Resource;
import org.springframework.stereotype.Controller;
import org.springframework.web.bind.annotation.RequestMapping;
import com.mvc.model.User;
import com.mvc.service.UserService;
@Controller("userController")
@RequestMapping("/user.do")
public class UserController {
    @Resource
    private UserService userService;
    @RequestMapping(params = "method=reg")
    public String reg(User user) {
        System.out.println("用户注册");
        if (userService.exist(user)) {
            return "reg_fail";
        }
        userService.add(user);
        return "reg_success";
    }
    @RequestMapping(params = "method=log")
```

```
    public String log(User user) {
        System.out.println("用户登录");
        if (userService.login(user)) {
            return "log_success";
        }
        return "log_fail";
    }
}
```

(5) View 层

在 WebContent 中新建两个“.jsp”文件，分别是“reg.jsp”文件和“log.jsp”文件。

①“Reg.jsp”文件的内容如下：

```
<%@ page language="java" contentType="text/html; charset=UTF-8"
    pageEncoding="UTF-8"%>
<!DOCTYPE html PUBLIC "-//W3C//DTD HTML 4.01 Transitional//EN" "http://www.w3.
org/TR/html4/loose.dtd">
<html>
<head>
<meta http-equiv="Content-Type" content="text/html; charset=UTF-8">
<title>Insert title here</title>
</head>
<body>
    <form action="user.do">
        <input type="text" name="name" />
        <br/>
        <input type="password" name="password" />
        <input type="hidden" name="method"value="reg" />
        <br/>
        <input type="submit" value="注册" />
    </form>
    <a href="/Spring MVC_Demo01/log.jsp">登录</a>
</body>
</html>
```

②“Log.jsp”文件的内容如下：

```
<%@ page language="java" contentType="text/html; charset=UTF-8"
    pageEncoding="UTF-8"%>
<!DOCTYPE html PUBLIC "-//W3C//DTD HTML 4.01 Transitional//EN"
"http://www.w3.org/TR/html4/loose.dtd">
<html>
<head>
<meta http-equiv="Content-Type" content="text/html; charset=UTF-8">
<title>Insert title here</title>
```

```
</head>
<body>
    <form action="user.do">
        <input type="text" name="name" />
        <br/>
        <input type="password" name="password" />
        <input type="hidden" name="method"value="log" />
        <br/>
        <input type="submit" value="登录" />
    </form>

    <a href="/Spring MVC_Demo01/reg.jsp">注册</a>
</body>
</html>
```

在/web-inf 文件夹中新建".jsp"文件夹创建四个".jsp"文件，分别为：reg_fail、reg_success、log_fail、log_success。四个文件中只做一些简单的提示。

③ Log_fail.jsp 文件内容如下：

```
<%@ page language="java" contentType="text/html; charset=UTF-8"
    pageEncoding="UTF-8"%>
<! DOCTYPE html PUBLIC "-//W3C//DTD HTML 4.01 Transitional//EN" "http://www.w3.
org/TR/html4/loose.dtd">
<html>
<head>
<meta http-equiv="Content-Type" content="text/html; charset=UTF-8">
<title>Insert title here</title>
</head>
<body>
    <h1>登录失败！用户名或密码错误</h1>
    <a href="/Spring MVC_Demo01/log.jsp">跳到登录页</a>
</body>
</html>
```

其他三个只改 body 中的提示内容，在此不再复述。

第四步：测试。

(1) 开启服务器

打开 tomcat 服务器，如图 7-30 所示。

图 7-30　Tomcat 服务器

(2) 在浏览器中浏览

右击 reg. jsp 文件选择 Run As On Server，如图 7－31 所示。

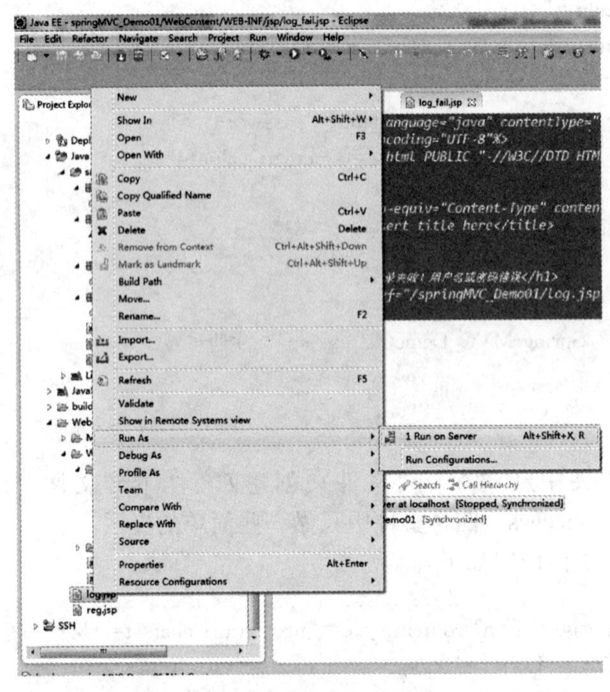

图 7－31　Tomcat 服务器运行

(3) 运行结果如图 7－32 所示。

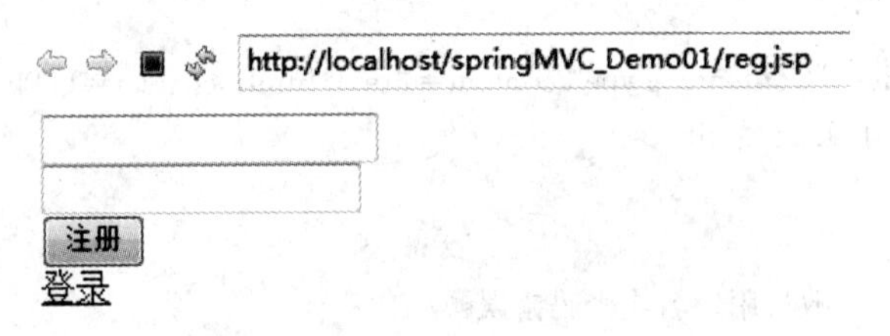

图 7－32　运行结果

巩固练习

1. 根据项目需求分组完成项目的其他功能。

附录 A

附录 A.1 测试用例说明文档

1. 为方便用例编写以及向 GT3K 导入，编写用例时，统一使用此模板，写作要求如下

层次	表示华为 GT3K 树状管理的层次，一个点表示一级，注意不要形成混乱(TD 中无要求根据华为方要求)。
测试项目	根据 TD 测试树集中 Subject 目录节点填写。注意：如果有多级目录必须采用如下格式"目录 1\目录 2\...\目录 N"。(在 GT3K 中，测试项目就是一个目录节点)
用例标题	在 TD 或 GT3K 中，对应具体的用例，显示在树状列表中，在该行中，测试项目列应该留空，在写用例时，该列应该明确表示出用例测试的内容，要知道与其他用例的差别，使用语言尽量简练，不要过长。
用例 ID	根据用例 ID 命名规则进行命名，避免重复。
用例级别	Level 1：基本。该类用例涉及系统基本功能，用于版本提交时作为"版本通过准则"。如存在不通过的项目时可考虑重新提交版本，例如通话不计费等。1 级用例的数量应受到控制。 Level 2：重要。2 级测试用例在非回归的系统测试版本中基本上都需要进行验证，以保证系统所有的重要功能都能够正常实现。在测试过程中可以根据版本当前的具体情况安排是否进行测试。 Level 3：一般。3 级测试用例使用频率较二级测试用例低，在非回归的系统测试版本中不一定都进行验证，而且在系统测试的中后期并不一定需要每个版本都进行测试。 Level 4：生僻。该类用例对应较生僻的预置条件和数据设置。虽然某些测试用例发现过较严重的错误，但是那些用例的触发条件非常特殊，仍然应该被置入 4 级用例中。有关用户界面的优化等方面的测试可归入 4 级用例。该类在实际使用中使用频率非常低、对用户可有可无。
前置条件	执行本测试用例前，被测试对象所需要具备的预置数据、所处状态或入口条件等要求。
用例描述	列出执行本测试用例所需的具体每一个输入或操作步骤。某些输入值可能要由下述的方式描述：如，允许一定范围的输入数值要指明公差值，引用常量表或事务文件的数据要指明引用的表或文件的名称。同时，如果输入值用到了相关的数据库、文件、终端显示信息、内存存储区数据以及操作系统底层传递的数值等，也要在此章节指明。 另外，如果有必要，也要说明各输入数值之间的依赖关系(例如，输入时间先后关系)。 对于集成测试和单元级别的测试用例，建议"输入"和"预期输出"应细化和具体到变量、输入点或检查点的具体数值，以使测试输入无二义性和测试检查结果唯一性，而系统测试因一些测试点的描述相对易于理解，可以不必给出具体的输入值和预期输出数值。
预期结果	描述被测项目或被测特性所希望或要求达到的输出(值)或指标(例如，系统响应时间)。 列出所有预期指标要求下的具体预期输出数值，如需要，也要列出描述预期输出数值的允许范围的公差值。

续表

测试类型	对应 TD 中几种测试类型,一般来说选择 MANUAL(手动测试)。
设计人	需要填写测试用例的设计者,方便以后用例跟踪。
状态	Design(设计),Imported(覆盖需求),Ready(预备),Repair(重新测试)。
测试者	执行测试用例的人员,若有多个人员用逗号","隔开。
Test Record:Fault Id	可以填写多个缺陷 ID,中间须用逗号","隔开。
Cycle N	每轮测试的结果记录,测试轮次根据需要可增删。 ① Pass:表示测试用例执行通过。 ② Fail:表示测试用例执行失败。 ③ Block:表示测试用例执行阻塞。 ④ Cancel:表示测试用例因为某种原因被取消执行。
备注	需要特别说明的地方,例如参考的 CheckList、文档、测试执行规程等。

2. 用例导入 TD 的方法

Step 1	安装 TD 导入工具 在服务上找到"TDMSExcelAddin. exe"工具安装,安装后重新打开 Excel。
Step 2	导入 GT3K GT3K 上菜单:"Tools"－＞"Export to TestDirector",在弹出的对话框中,填写 TD 服务器的地址:http://172.16.1.192/tdbin,选择"域名"和"项目名",输入在该项目中的"用户名"和"密码"在弹出的对话框中制定各自段与 TD 中各字段的关联关系。TD 的项对应 Excel 的列。

3. 用例导入 GT3K 的方法

Step 1	转成 XML 文件 GT3K 上菜单:"Tools"－＞"Converting Tools"－＞"Excel2Xml",在弹出的对话框中,选择文件、表格及输出文件名。
Step 2	导入 GT3K 首先选定导入的目的节点,点击 GT3K 上菜单:"Import File",选择文件后在弹出的对话框中制定各自段与 GT3K 中各字段的关联关系。
注意	如果是回归版本新增用例,请选择对应的新版本节点再导入,这样便于计算新增用例数量。

附录 A.2 [STORY.eTraining.001.01]—[需求库管理]

1. Introduction 介绍

培训需求的新增、修改、删除、查询、详情、导出等功能。

2. Story 开发设计

(1) 页面原型设计(可选)

功能点完成之后提供

(2) Functions 功能描述

(3) Design Description 模块设计描述

流程图,逻辑图

(4) Function Illustration 功能实现说明

① 涉及影响的功能点

② 修改文件的列表:

java : src/com /common/dao/ * * *.java

jsp: WebRoot/WEB-INF/jsp/agt/common/ * * *.jsp

js: WebRoot/js/agt/exam/ * * *.js

(5) Database Design(Optional)数据库设计(可选)

① T_AGT_TRAINDEMAND 培训需求

字段名	字段类型	缺省值	中文显示名称	约束关系	意义
TRAINDEMANDID	VARCHAR2(20)		培训需求 ID	非空	主键
ORIGIN	VARCHAR2(1)		来源	非空	0:直接创建 1:效能分析
INSTANCYDEGREE	VARCHAR2(1)		紧急程度	非空	0:一般 1:紧急 2:特急
LESSONID	VARCHAR2(20)		培训课程 ID		参照 T_AGT_LESSON 表的 LESSONID 字段
TITLE	VARCHAR2(100)		课程名称		参照 T_AGT_LESSON 表的 TITLE 字段
DESCRIPTION	VARCHAR2(1000)		需求说明		新增培训需求时该字段与培训课程 ID 字段不能同时为空
STARTTIME	DATE		开始时间		建议培训开始时间
ENDTIME	DATE		结束时间		建议培训结束时间

续表

字段名	字段类型	缺省值	中文显示名称	约束关系	意义
TARGET	VARCHAR2(1000)		培训目标		
STATE	VARCHAR2(1)		状态		0:预处理 1:未处理 2:已处理 3:退回
CREATOR	VARCHAR2(20)		创建人		参照 T_UCP_STAFFBASICINFO 的 STAFFID 字段
CREATETIME	DATE		创建时间		根据时间分区
QUESTIONID	VARCHAR2(20)		问题 ID		
REJECTREASON	VARCHAR2(1000)		驳回原因		当培训需求为非退回状态时该字段为空
TRAINPLANID	VARCHAR2(20)		培训 ID		参照 T_AGT_TRAINPLAN 表的 TRAINPLANID 字段

说明:记录培训的需求信息。

主键:PK_T_AGT_TRAINDEMAND(TRAINDEMANDID)。

数据来源:

a. 数据来源于问题库系统,则来源字段值为效能分析;

b. 数据来源于培训需求新增页面,则来源字段值为直接创建。

② T_AGT_TRAINDEMANDSTAFF 培训需求对象

字段名	字段类型	缺省值	中文显示名称	约束关系	意义
TRAINDEMANDID	VARCHAR2(20)		培训需求 ID	非空	参照 T_AGT_TRAINDEMAND 表的 TRAINDEMANDID 字段
STAFFID	VARCHAR2(20)		培训对象	非空	参照 T_UCP_STAFFBASICINFO 的 STAFFID 字段

说明:记录欲指定为培训参与人员的对象。

主键:PK_T_AGT_TRAINDEMANDSTAFF(TRAINDEMANDID+STAFFID)。

来源:培训新增、修改页面指定。

③ T_AGT_TRAINDEMANDMANAGEFLOW 培训需求处理流程

字段名	字段类型	缺省值	中文显示名称	约束关系	意义
TRAINDEMANDID	VARCHAR2(20)		培训需求 ID	非空	参照 T_AGT_TRAINDEMAND 表的 TRAINDEMANDID 字段
MODIFIER	VARCHAR2(20)		修改人	非空	参照 T_UCP_STAFFBASICINFO 的 STAFFID 字段
MODIFYTIME	DATE		修改时间	非空	根据时间分区
MODSTATE	VARCHAR2(1)		修改后状态	非空	0:预处理 1:未处理 2:已处理 3:退回

说明:培训需求在新增、修改之后均会在该表新增记录。

(6) Internal Dependency Description 依赖性描述(可选)

N/A

(7) Performance Requirements 性能需求和兼容性(可选)

N/A

3. Story 测试设计

(1) 测试方案说明

这里需要描述清楚 story 的测试范围,测试重点,从 story 使用场景的角度分析,需要进行哪些测试,如 ST,SDV,是否可以进行自动化,是否需要测试装。

(2) 测试功能列表

(3) Story 验收标准

① Story 修改描述

修改文件说明				
No.	修改文件	路径	类别	修改说明

② Story 资料建议

附录 B

附录 B.1 教育资源平台数据字典

日志表

字段名称	字段含义	数据类型	宽度	NULL	注
ID	表的主键	NUMBER		N	记录的唯一标识
TUS_ID	用户 ID	NUMBER		Y	关联用户表
RES_ID	资源 ID	NUMBER		Y	关联资源表
TIME	操作时间	VARCHAR2		Y	
IPADDR	IP 地址	VARCHAR2		Y	
OPERATETYPE	操作类型	VARCHAR2		Y	
KEYWORDS	关键字	VARCHAR2		Y	
REMARKS	备注信息	VARCHAR2		Y	
USETIME	花费时间	LONG		Y	查看执行效率
RES_TITLE	资源标题	VARCHAR2		Y	

书签表

字段名称	字段含义	数据类型	宽度	NULL	注
ID	表的主键	NUMBER		N	记录的唯一标识
NAME	书签名称	VARCHAR2		N	
PREVIOUSPAGE	当前页数	NUMBER		Y	
TOTALPAGE	总页数	NUMBER		Y	
RESOURCEID	资源 ID	NUMBER		Y	
CREATEDATE	创建时间	DATE		Y	
USERID	创建用户 ID	NUMBER		Y	

资源分类表

字段名称	字段含义	数据类型	宽度	NULL	注
ID	表的主键	NUMBER	12	N	记录的唯一标识
CODE	分类代码	VARCHAR2	100	Y	
CETEGORYNAME	分类名称	VARCHAR2	100	Y	
PARENTID	父节点 ID	NUMBER	12	Y	
RANK	排名	NUMBER	12	Y	
STATUS	开启状态	NUMBER	12	Y	

操作类型表

字段名称	字段含义	数据类型	宽度	NULL	注
ID	表的主键	NUMBER	12	N	记录的唯一标识
METHOD	方法名称	VARCHAR2	100	Y	
METHODNAME	方法意义	VARCHAR2	100	Y	
IDPOSITION	ID 所在位置	VARCHAR2	12	Y	

资源包表

字段名称	字段含义	数据类型	宽度	NULL	注
ID	表的主键	NUMBER	6	N	记录的唯一标识
PACKAGENAME	资源包名称	VARCHAR2	60	N	
SUMMARIZE	资源包大小	VARCHAR2	200	Y	
STATUS	资源包状态	NUMBER	6	Y	
USERID	创建用户	NUMBER	6	Y	
CREATEDATE	创建时间	DATE		Y	
CATEGORYID	资源包类型	NUMBER	6	Y	

资源栏目表

字段名称	字段含义	数据类型	宽度	NULL	注
ID	表的主键	NUMBER	16	N	记录的唯一标识
COLUMNNAME	栏目名称	VARCHAR2	60	N	
PARENTID	父级栏目	NUMBER	16	Y	
STATUS	栏目状态	NUMBER	6	Y	
BUILDDATE	创建时间	DATE		Y	
RANK	排名	NUMBER	6	Y	
URL	路径	VARCHAR2	100	Y	
PACKAGEID	资源包 ID	NUMBER	6	Y	

资源表

字段名称	字段含义	数据类型	宽度	NULL	注
ID	表的主键	NUMBER	16	N	记录的唯一标识
RESID	资源标识符	VARCHAR2	64	N	
TITLE	资源名称	VARCHAR2	1000	N	
RESLANGUAGE	语种	NUMBER		Y	
DESCRIPTION	描述	VARCHAR2	2000	Y	
KEYWORD	关键字	VARCHAR2	100	Y	
FORMAT	格式	NUMBER		Y	
SUBJECT	学科门类	NUMBER		Y	
PERIOD	时段	NUMBER		Y	
TYPE	资源类别	NUMBER		Y	
AUTHOR	创建人	VARCHAR2	200	Y	
CREATETIME	创建时间	VARCHAR2	200	Y	
VERSION	版本	VARCHAR2	200	Y	
RESSIZE	资源大小	FLOAT		Y	
LOCATION	位置	VARCHAR2	1000	Y	
INTENDED_END_USRER_ROLE	浏览次数	NUMBER		Y	
LEARNINGTIME	下载次数	NUMBER		Y	
COPYRIGHT_RESTRICTIONS	版权限制	VARCHAR2	60	Y	
COPYRIGTH_OWNER	版权所有	VARCHAR2	200	Y	
COST	是否收费	NUMBER		Y	
RESSTATE	资源状态	NUMBER		Y	
EXPENSE	费用	NUMBER		Y	
DOWNLOAD	是否可下载	NUMBER		Y	
TERM	所属学期	NUMBER		Y	
REASON	未通过原因	VARCHAR2	500	Y	

代码集表

字段名称	字段含义	数据类型	宽度	NULL	注
ID	表的主键	NUMBER	12	N	记录的唯一标识
DICTCODE	编码	VARCHAR2	100	Y	
DICTNAME	名称	VARCHAR2	100	Y	
DICTGROUPID	分组序号	NUMBER	12	N	
REMARK	备注	VARCHAR2	100	Y	
STATUS	状态	NUMBER	6	Y	

代码集分组表

字段名称	字段含义	数据类型	宽度	NULL	注
ID	表的主键	NUMBER	12	N	记录的唯一标识
GROUPCODE	编码	VARCHAR2	100	Y	
GROUPNAME	名称	VARCHAR2	100	Y	
STATUS	状态	NUMBER	6	Y	

代码集子类表

字段名称	字段含义	数据类型	宽度	NULL	注
ID	表的主键	NUMBER	12	N	记录的唯一标识
ITEMCODE	编码	VARCHAR2	100	Y	
ITEMNAME	名称	VARCHAR2	100	Y	
ITEMVALUE	代码子集值	VARCHAR2	100	Y	
DICTID	代码集 ID	NUMBER	12	N	
STATUS	状态	NUMBER	12	Y	

友情链接表

字段名称	字段含义	数据类型	宽度	NULL	注
ID	表的主键	NUMBER	6	N	记录的唯一标识
LINKNAME	名称	VARCHAR2	100	Y	
LINKICON	图标	VARCHAR2	100	Y	
LINKADDRESS	地址	VARCHAR2	100	N	
RANK	排名	NUMBER	6	Y	

菜单表

字段名称	字段含义	数据类型	宽度	NULL	注
ID	表的主键	NUMBER	12	N	记录的唯一标识
CODE	菜单编号	VARCHAR2	60	N	
NAME	菜单名称	VARCHAR2	100	N	
ICON	菜单图标	VARCHAR2	60	Y	
URL	菜单地址	VARCHAR2	300	Y	
PARENTID	父节点	NUMBER	12	Y	
RANK	排序	NUMBER	12	Y	
STATUS	状态	NUMBER	12	Y	

新闻表

字段名称	字段含义	数据类型	宽度	NULL	注
ID	表的主键	NUMBER	12	N	记录的唯一标识
TITLE	新闻标题	VARCHAR2	120	N	
NEWSCONTENT	新闻内容	VARCHAR2	4000	Y	
NEWSCATEGORY	新闻分类	NUMBER	12	N	
PUBTIME	发布时间	DATE		Y	
PUBLISHER	发布人	NUMBER	12	N	
PUBORG	发布机构	NUMBER	12	Y	
STATUS	状态	NUMBER	5	Y	
ISTOP	置顶	NUMBER	5	Y	

新闻分类表

字段名称	字段含义	数据类型	宽度	NULL	注
ID	表的主键	NUMBER	12	N	记录的唯一标识
CATEGORYNAME	分类名称	VARCHAR2	60	N	
ISCOMMENT	是否需要评论	NUMBER	6	Y	
ISANONYMOUS	是否匿名	NUMBER	6	Y	

权限表

字段名称	字段含义	数据类型	宽度	NULL	注
ID	表的主键	NUMBER	12	N	记录的唯一标识
OPERATECODE	权限代码	VARCHAR2	200	Y	
OPERATENAME	名称	VARCHAR2	100	Y	
PARENTID	父节点	NUMBER	12	Y	
URL	链接	VARCHAR2	300	Y	
RANK	排序号	NUMBER	12	Y	
TYPE	类型	NUMBER	12	Y	

组织机构表

字段名称	字段含义	数据类型	宽度	NULL	注
ID	表的主键	NUMBER	12	N	记录的唯一标识
ORGNAME	机构名称	VARCHAR2	100	N	
ORGCODE	机构编号	VARCHAR2	60	N	
PARENTID	父节点	NUMBER	12	Y	
CREATEDATE	创建时间	DATE		Y	
STATUS	状态	NUMBER	12	Y	
RANK	排序	NUMBER	22	Y	

用户注册表

字段名称	字段含义	数据类型	宽度	NULL	注
ID	表的主键	NUMBER	12	N	记录的唯一标识
EMAIL	邮箱	VARCHAR2	60	Y	
USERNAME	真实姓名	VARCHAR2	60	N	
PASSWORD	密码	VARCHAR2	60	N	
CREATEDATE	注册时间	DATE		Y	
REMARK	备注	VARCHAR2	60	Y	
ACTIVE	是否激活	NUMBER	6	N	
USERID	用户名	VARCHAR2	60	Y	
SUBJECT	学科	VARCHAR2	60	Y	
GRADE	年级	VARCHAR2	60	Y	
USERTYPE	类型	VARCHAR2	60	Y	
PERIOD	学段	VARCHAR2	60	Y	

角色表

字段名称	字段含义	数据类型	宽度	NULL	注
ID	表的主键	NUMBER	12	N	记录的唯一标识
ROLECODE	角色编号	VARCHAR2	100	Y	
ROLENAME	角色名称	VARCHAR2	100	Y	
CATEGORY	角色类型	NUMBER		Y	

角色类型表

字段名称	字段含义	数据类型	宽度	NULL	注
ID	表的主键	NUMBER	12	N	记录的唯一标识
CATEGORYNO	分类编号	VARCHAR2	60	Y	
CATEGORYNAME	分类名称	VARCHAR2	60	Y	
CATEGORYTYPE	分类类型	NUMBER		Y	

角色权限表

字段名称	字段含义	数据类型	宽度	NULL	注
ID	表的主键	NUMBER	12	N	记录的唯一标识
ROLEID	角色编号	NUMBER		Y	
OPERATEID	权限编号	NUMBER		Y	
AUDITFLAG	审批标志	NUMBER		Y	

学科年级关联表

字段名称	字段含义	数据类型	宽度	NULL	注
ID	表的主键	NUMBER	12	N	记录的唯一标识
PERIOD_ID	学段编号	NUMBER	12		
SUBJECT_ID	学科编号	NUMBER	12		

系统参数表

字段名称	字段含义	数据类型	宽度	NULL	注
ID	表的主键	NUMBER	12	N	记录的唯一标识
PARAMNAME	参数名称	VARCHAR2	60		
PARAMVALUE	参数值	VARCHAR2	200		
REMARK	备注	VARCHAR2	300		

用户表

字段名称	字段含义	数据类型	宽度	NULL	注
ID	表的主键	NUMBER	12	N	记录的唯一标识
USERID	用户名	VARCHAR2	60	Y	
USERNAME	真实姓名	VARCHAR2	60	Y	
PASSWORD	密码	VARCHAR2	60	Y	
USERTYPE	用户类型	NUMBER	6	Y	
LOCKED	是否锁定	NUMBER	6	Y	
CREDIT	积分	NUMBER	6	Y	
ORGID	隶属组织	NUMBER	12	Y	
PHOTO	头像	VARCHAR2	60	Y	
CREATEDATE	注册时间	DATE		Y	
EMAIL	邮箱	VARCHAR2	60	Y	
SUBJECT	学科	VARCHAR2	60	Y	
GRADE	年级	VARCHAR2	60	Y	
PERIOD	学段	VARCHAR2	60	Y	

用户兴趣表

字段名称	字段含义	数据类型	宽度	NULL	注
ID	表的主键	NUMBER	12	N	记录的唯一标识
USERID	用户编号	NUMBER		Y	
INTEREST	兴趣	VARCHAR2	200	Y	
GOODAT	擅长	VARCHAR2	200	Y	

用户信息表

字段名称	字段含义	数据类型	宽度	NULL	注
ID	表的主键	NUMBER	12	N	记录的唯一标识
USERID	用户编号	NUMBER	12	N	
SEX	性别	VARCHAR2	60	Y	
BIRTHDAY	生日	VARCHAR2	200	Y	
CONSTELLATION	星座	VARCHAR2	60	Y	
VOCATIONAL	职业	VARCHAR2	60	Y	
PROFESSIONAL	专业	VARCHAR2	60	Y	
EDUCATION	教育背景	VARCHAR2	60	Y	
INTEREST	兴趣	VARCHAR2	60	Y	
LIVEADD	居住地址	VARCHAR2	2060	Y	
HOMEADD	家庭地址	VARCHAR2	60	Y	
MARRIAGE	婚姻状况	VARCHAR2	60	Y	
BLOOD	血型	VARCHAR2	60	Y	
SCHOOL	学校	VARCHAR2	60	Y	
YEAR	入学时间	VARCHAR2	200	Y	
SCHOOLADD	学校地址	VARCHAR2	60	Y	
COMPANY	所在公司	VARCHAR2	60	Y	
STARTTIME	入职时间	VARCHAR2	200	Y	
COMPANYADD	公司地址	VARCHAR2	60	Y	
TELPHONE	电话号码	VARCHAR2	60	Y	
EMAIL	邮箱	VARCHAR2	60	Y	
QQ	QQ 号码	VARCHAR2	60	Y	
HEAD	用户头像	VARCHAR2	60	Y	

用户排名表

字段名称	字段含义	数据类型	宽度	NULL	注
ID	表的主键	NUMBER	12	N	记录的唯一标识
USERID	用户编号	NUMBER	16	N	
INTEGRAL	积分	NUMBER	16	Y	
EXPERIENCE	经验值	NUMBER	16	Y	
URESOURCENUM	上传数量	NUMBER	16	Y	
RANK	级别	NUMBER	6	Y	
DRESOURCENUM	下载数量	NUMBER	16	Y	

用户角色表

字段名称	字段含义	数据类型	宽度	NULL	注
ID	表的主键	NUMBER	12	N	记录的唯一标识
USERID	用户编号	NUMBER		Y	
ROLEID	角色编号	NUMBER		Y	
ORGID	隶属机构	NUMBER		Y	
AUDITFLAG	审批标题	NUMBER		Y	